AF324853

Quasi-Orthogonal Space-Time Block Code

Communications and Signal Processing

Editors: Prof. A. Manikas & Prof. A. G. Constantinides
(Imperial College London, UK)

Communications and Signal Processing – Vol. 2

Quasi-Orthogonal Space-Time Block Code

Chau Yuen

Institute for Infocomm Research, Singapore

Yong Liang Guan

Nanyang Technological University, Singapore

Tjeng Thiang Tjhung

Institute for Infocomm Research, Singapore

Imperial College Press

Published by

Imperial College Press
57 Shelton Street
Covent Garden
London WC2H 9HE

Distributed by

World Scientific Publishing Co. Pte. Ltd.

5 Toh Tuck Link, Singapore 596224

USA office: 27 Warren Street, Suite 401-402, Hackensack, NJ 07601

UK office: 57 Shelton Street, Covent Garden, London WC2H 9HE

British Library Cataloguing-in-Publication Data
A catalogue record for this book is available from the British Library.

Communications and Signal Processing — Vol. 2
QUASI-ORTHOGONAL SPACE-TIME BLOCK CODE

ISBN-13 978-1-86094-868-8
ISBN-10 1-86094-868-5

Desk editor: Tjan Kwang Wei

Printed by FuIsland Offset Printing (S) Pte Ltd, Singapore

Foreword

Owing to their ability to multiply wireless capacity, mitigate slow fading, and facilitate new adaptive communications beyond the limits of conventional single-antenna wireless systems, MIMO and Space Time Coding techniques (which combine coding, modulation and signal processing designs in systems employing multiple transmit/receive antennas) have generated much research interest in recent years. Their adoption in cellular mobile radio, wireless LAN and wireless MAN standards have also marked their increasing significance in commercial broadband wireless systems.

An important class of space-time code is the Orthogonal Space-Time Block Code (O-STBC), which is attractive for its low decoding complexity, low decoding latency, and ability to provide full transmit diversity for mitigating slow fading by requiring multiple antennas only at the base stations or access points (hence low cost per user). However O-STBC suffers from low code rate (hence non-optimal MIMO capacity) when used with more than 2 transmit antennas and complex modulation. This rate disadvantage can be alleviated by a class of group-decodable STBC design called the Quasi-Orthogonal STBC (QO-STBC). The decoding complexity of QO-STBC is higher than that of O-STBC, but it can be capped by proper code design.

This monograph focuses on the design and analysis of full-diversity QO-STBC with very low decoding complexity and high code rate. Much of the monograph is dedicated to double-symbol-decodable QO-STBC and single-symbol-decodable QO-STBC, which have the two lowest levels of decoding complexity among all QO-STBC. Apart from coherent code detection, differential space-time modulation designs for the non-coherent (blind) detection of double- and single-symbol-decodable QO-STBC are also presented. The latter part of this monograph shifts the design focus away from minimum decoding complexity to constant full rate (i.e. code rate = 1) for arbitrary number of transmit antennas, and finally to code rate > 1.

Chapter 1 of the monograph sets out the essential foundations for subsequent technical discussions. They include the MIMO channel model, the dispersion matrix representation of QO-STBC, the concept of equivalent channel matrix, as well as the code parameters, performance measures, and the well-known *Rank and Determinant* design criteria of QO-STBC. Chapter 2 reviews the state of the arts in O-STBC and QO-STBC, and compares their essential differences in code rate, decoding complexity and number of antennas supported. The classical Amicable Orthogonal Design (AOD) for constructing O-STBC and constellation rotation (CR) technique for rendering full diversity in QO-STBC are also discussed. Chapter 3 elaborates on the concept of group-wise symbol decoding that gives QO-STBC its low decoding complexity advantage. The constraints required of the QO-STBC dispersion matrices to render it group-decodable are derived. They are shown to lead to the unexpected finding that the classical CR technique applied on many past QO-STBC actually leads to increased decoding complexity for the resultant codes. Hence an alternative constellation transformation technique, called Group-Constrained Linear Transformation (GCLT), is presented to resolve this issue. Chapter 4 moves the low decoding complexity code design attempt a step further by introducing a class of single-symbol-decodable QO-STBC called Minimum Decoding Complexity QO-STBC (MDC-QOSTBC), which has the lowest possible decoding complexity among any QO-STBC. Its dispersion matrix constraints, code construction rules, optimum code parameters, code performance, and antenna downscaling methods are discussed in depth. A new AOD concept called Preferred AOD Pair (developed from the classical AOD discussed in Chapter 2) is also presented to derive the maximum achievable code rates of MDC-QOSTBC for any given number of transmit antennas. In Chapter 5 we switch our focus to the design of non-coherent detection for QO-STBC with low decoding complexity. In contrast to existing works, we obtain non-coherent QO-STBC by designing special "joint constellation" sets for the double- and single-symbol-decodable QO-STBC (discussed in Chapter 3 and 4) to achieve certain unitary or quasi-unitary code structure without affecting the quasi-orthogonality of the code. To our knowledge, the single-symbol-decodable non-coherent QO-STBC pioneered by us was the first and remained the lowest decoding complexity design with higher code rate than non-coherent O-STBC. In Chapter 6 we move the code design emphasis away from low decoding complexity to flexible rate codes. Two new classes of STBC are presented: constant full-rate QO-STBC with 4 separately-decodable symbol groups for arbitrary number of transmit antennas, and QO-STBC with code rate > 1. The code search methodology of the latter is based on an interesting application of the QO-STBC code constraints derived in Chapter 3 into the classical graph theory and tree search algorithms. Among the new codes found are the first-ever QO-STBC's with code rate of 5/4 for 4 transmit antennas. Finally, in Chapter 7 we outline the potential roles of QO-STBC in large-scale communication systems with hybrid ARQ, FEC, super space-time trellis coding or MIMO OFDM, as well as in commercial wireless standards such as 3GPP LTE.

We thank the Imperial College Press for initiating this monograph idea. We hope that the monograph will give the readers a comprehensive and integrated picture of the many recent advances in this interesting class of space-time code, and spur further innovations to unlock the full potential of MIMO communication for realizing mankind's dream of the Wireless Utopia.

Chau <u>Yuen</u>
Yong Liang <u>Guan</u>
Tjeng Thiang <u>Tjhung</u>

Contents

Chapter 1

Introduction of MIMO Channel and Space-Time Block Code

In the past few years, there has been a phenomenal increase in consumers' as well as manufacturers' interest in wireless communications. This is due to the advances of wireless communication technology providing the advantages of wide area coverage without wires, and most importantly, allowing mobility while communicating. Beyond the success of the established technologies such as mobile telephony, a wide range of new wireless communications services are being developed. For example, there has been growing interest in providing broadband wireless Internet services with rich multimedia contents at near wire-line data rates. However, the wireless channel suffers from random signal attenuation and phase distortion due to the destructive superposition of multiple received signals in a multipath propagation environment, a phenomenon commonly called *fading*. To mitigate fading and push the capacity of wireless channel to a higher limit, the use of multiple transmitting and/or receiving antennas, or the so-called multiple-input multiple-output (MIMO) concept, has recently been proposed.

1.1 MIMO Channel for Wireless Communications

Fading makes it extremely difficult for the receiver to recover the transmitted signal unless the receiver is provided with some form of *diversity*, i.e. replicas of the same transmitted signal with uncorrelated attenuation. In fact, diversity combining technology has been one of the most important contributors to reliable wireless communications. Ways to achieve diversity include:

- *Temporal Diversity*: In this scheme, channel coding in conjunction with time interleaving is used. Thus replicas of the transmitted signal are provided to the receiver in the form of redundancy in the temporal domain. However, in slow fading channels, temporal diversity is not an option for delay-sensitive applications.
- *Frequency Diversity*: In this scheme, the fact that signals that are transmitted on different frequencies tend to experience different fading effects is exploited. Thus replicas of the transmitted signal are provided to the receiver in the form of redundancy in the frequency domain. However, this scheme is not bandwidth-efficient.
- *Spatial Diversity*: In this scheme, spatially separated antennas are used to provide diversity in the spatial domain. Diversity combining technique is then used to select or combine the signals that have been transmitted or received on different antennas.

Spatial diversity is attractive as diversity can be obtained with no penalty in bandwidth efficiency. It can be implemented by deploying multiple antennas at the transmitter and/or the receiver. Depending on the location of the antennas, we can classify wireless communication system employing spatial diversity into the following three configurations:

- *Single Input Multiple Output* (SIMO): When there are single transmit antenna but multiple receive antennas, i.e. *receive diversity*.
- *Multiple Input Single Output* (MISO): When there are multiple transmit antennas but one receive antenna, i.e. *transmit diversity*.
- *Multiple Input Multiple Output* (MIMO): When there are multiple transmit antennas and multiple receive antennas, i.e. both transmit and receive diversity are used.

Besides providing spatial diversity, it has been shown in [1,2] that the capacity of a wireless channel grows linearly with the number of transmit and receive antennas, hence a MIMO system can be used to boost the capacity of wireless channel too.

Considering the fact that mobile receivers are typically required to be small and cost-effective, it may not be practical to deploy receive diversity at the mobile terminal. This motivates many researchers to consider transmit diversity by deploying multiple antennas at the base station. Moreover, in economic terms, the cost of multiple transmit antennas at the base station can be amortized over numerous mobile users. Hence transmit diversity has been identified as one of the key contributing technologies to the downlinks of 3G wireless systems such as W-CDMA and CDMA2000 [3]. There are generally three categories of transmit diversity:

- *Feedback Scheme*: This involves the feedback of channel state information (CSI, typically including channel gain and phase information) from the receiver to the transmitter in order to adapt the transmitter to the channel during the next transmission epochs. It is also commonly known as the "closed-loop" system.
- *Feedforward Scheme*: This involves the receiver making use of feedforward information sent by the transmitter, such as pilot symbols, to estimate the channel, but no channel feedback information is sent back to the transmitter. It is also commonly known as the "open-loop" or "coherent" system.
- *Blind Scheme*: This requires no feedback of CSI or feedforward of pilots, and the receiver simply makes use of the received signal to attempt data recovery without the knowledge of CSI. It is also commonly known as the "non-coherent" system.

To demonstrate the benefit of transmit diversity under the feedforward scheme, the bit error rate (BER) performance versus bit energy to noise spectral density ratio (E_b/N_o) of a typical transmit diversity scheme with various number of transmit antennas and one receive antenna at a spectral efficiency of 2 bits/sec/Hz (bps/Hz) is illustrated in Fig. 1.1. These results are achieved by using *Space-Time*

Block Code (STBC), a type of feedforward transmit diversity coding scheme that will be the main focus of this monograph. It can be seen that when there is only one transmit antenna, more than 15dB increase in E_b/N_o is required to achieve a BER of 10^{-3} in a Rayleigh faded wireless channel over an additive white Gaussian noise (AWGN) channel. By employing multiple transmit antennas to provide transmit diversity, the BER can be significantly reduced, such that the BER curve decays faster with E_b/N_o. This is due to the multiple transmit antennas providing higher spatial diversity level. However, unlike receive diversity that can be achieved by simply performing the diversity combining at the receiver side, transmit diversity requires some form of signal processing, generally known as space time coding, on the transmitted signals in order to achieve signal enhancement at the receiver.

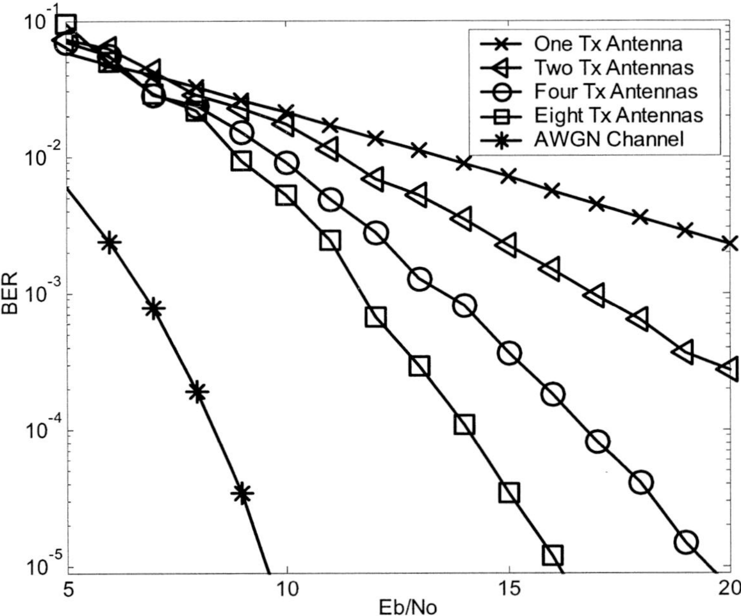

Fig. 1.1 Feedforward transmit diversity with various diversity levels.

Space-Time Coding (STC) is a technique that combines coding, modulation and signal processing to achieve transmit diversity. The first

STC proposed in the literature is *Space-Time Trellis Code* (STTC) [4], which has a good decoding performance but decoding complexity that increases exponentially with the transmission rate. In addressing the issue of decoding complexity of STTC, *Space-Time Block Code* (STBC) was subsequently proposed. Alamouti [5] discovered a remarkable STBC scheme for two transmit antennas. This scheme supports linear decoding complexity for maximum-likelihood (ML) decoding, which is much simpler than the decoding of STTC. It can achieve the same diversity gain as a corresponding STTC for two transmit antennas, though with a shortfall in coding gain. Despite the lower coding gain, Alamouti's scheme is very appealing in terms of implementation simplicity. Hence it motivates a search for similar schemes for more than two transmit antennas, to achieve diversity level higher than two. As a result, *Orthogonal Space-Time Block Code* (O-STBC) was introduced by Tarokh *et al.* in [6]. O-STBC is a generalization of the Alamouti's scheme to an arbitrary number of transmit antennas. It retains the property of having linear maximum-likelihood decoding with full transmit diversity.

Although O-STBC can provide full diversity at low computational cost, [7] showed that it suffers a loss in capacity when (1) there are multiple receive antennas, (2) the code rate is less than one. As rate-1 O-STBC with complex constellation is not possible for more than two transmit antennas [6], O-STBC design for more than two transmit antennas will always suffer capacity loss.

To address the issue of capacity loss, various non-orthogonal STBC designs have been proposed. An interesting one among them is the *Quasi-Orthogonal STBC* (QO-STBC) [8,9,10], which is designed to achieve a higher code rate than O-STBC by partially (instead of fully, as in the case of other non-orthogonal STBCs) relaxing the orthogonality of an O-STBC. For example, the ML decoding of the full-rate QO-STBC in [8] for four transmit antennas can be achieved by jointly detecting two out of four complex symbols in the codeword, and separately doing the same for the remaining two complex symbols. Due to this low decoding complexity advantage of QO-STBC, as well as its ability to achieve full transmit diversity, we seek to provide a complete study on QO-STBC in this monograph, and to seek further improvements in its design.

The focus of this monograph is on the spatial diversity for MISO or MIMO channel with feedforward and blind configurations by using QO-STBC in non-frequency selective channel and uncoded system. We focus on the fundamental code design issues of QO-STBC, as they serve as the basic element for extension to closed-loop MIMO system, coded MIMO system and MIMO systems for frequency selective fading channels, which will be briefly discussed in the last chapter of this monograph.

1.2 Transmit Diversity with Space-Time Block Code

Before we introduce the transmit diversity scheme based on Space-Time Block Code (STBC), we first review the traditional receive diversity scheme with *maximal ratio combining* (MRC) for one transmit antenna and two receive antennas, using Fig. 1.2 as an example.

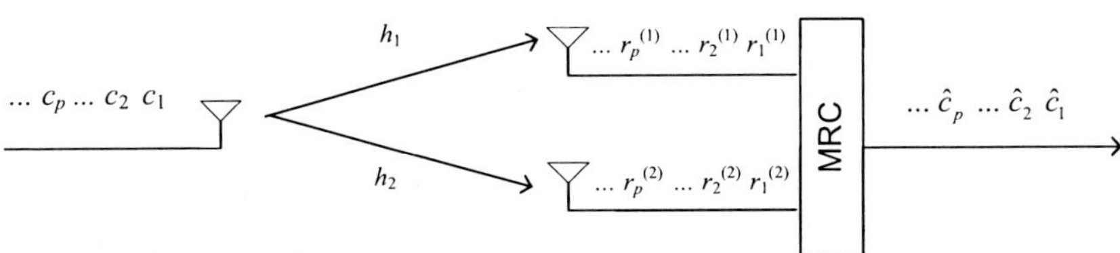

Fig. 1.2 Receive diversity with one transmit and two receive antennas

Denoting the transmitted signal at time p as c_p, and the received signals on the first and second receive antennas as $r_p^{(1)}$ and $r_p^{(2)}$ respectively, we obtain the following expressions:

$$r_p^{(1)} = h_1 c_p + \eta_p^{(1)},$$
$$r_p^{(2)} = h_2 c_p + \eta_p^{(2)}, \tag{1.1}$$

where h_1 and h_2 are the CSI or path gain from the transmit antenna to the first and second receive antennas respectively, and $\eta_p^{(1)}$ and $\eta_p^{(2)}$ are additive white Guassian noises (AWGN) at the respective received antennas at time instant p.

If only one receive antenna is available, the transmitted symbols can be estimated as follows, by assuming that the CSI is known accurately:

$$\hat{c}_p = h_i^* r_p^{(i)} \qquad\qquad i = 1, 2$$

$$= |h_i|^2 c_p + h_i^* \eta_p^{(i)},$$

(1.2)

where * denotes the complex conjugate and $|\,.\,|$ denotes the magnitude of a complex element.

When multiple receive antennas are available, to retrieve the data symbols utilizing the diversity signals provided by multiple receive antennas, we perform MRC as follows:

$$\hat{c}_p = h_1^* r_p^{(1)} + h_2^* r_p^{(2)}$$

$$= \left(|h_1|^2 + |h_2|^2 \right) c_p + h_1^* \eta_p^{(1)} + h_2^* \eta_p^{(2)}.$$

(1.3)

We can see from both equations (1.2) and (1.3) that the signal estimate $\hat{c}_p$ consists of the actual signal c_p weighted by a factor related to the fading magnitude, then summed with a noise term. We can say that (1.3) gives a better estimate than (1.2) because the chance that both h_1 and h_2 in (1.3) fade simultaneously is much smaller than the chance that h_1 in (1.2) fades. Statistically, if h_1 and h_2 are Rayleigh-distributed and uncorrelated, $|h_1|^2$ will have a Chi-Square distribution with two degrees of freedom, while $|h_1|^2 + |h_2|^2$ will have a Chi-Square distribution with four degree of freedom, hence a lower probability of deep fade. This explains the diversity gain of a MRC receive diversity scheme over a non-diversity scheme.

However, the deployment of multiple receive antennas at the mobile station may not be feasible due to size and cost constraints, this has therefore motivated the research of transmit diversity to provide spatial diversity for the downlink channel using multiple transmit antennas at the base station. In [5], Alamouti proposed a simple two-antenna transmit diversity scheme which up to today remains the only O-STBC that achieves the same diversity gain as the two-antenna receive MRC diversity scheme at full rate for any complex constellation.

The Alamouti O-STBC scheme is described as follows. Considering a system with two transmit antennas and one receive antenna as shown in Fig. 1.3, at a given symbol period, two signals are simultaneously transmitted from two antennas using the same bandwidth. At time $2p-1$, the signal c_{2p-1} is transmitted from the first antenna, while the signal c_{2p} is transmitted from the second antenna. During the next symbol period $2p$, signal $-c_{2p}^*$ is transmitted from the first antenna, and signal c_{2p-1}^* is transmitted from the second antenna. In order to normalize the total transmission power to be the same as the receive diversity scheme in Fig. 1.2, the transmission power from each transmit antenna is halved.

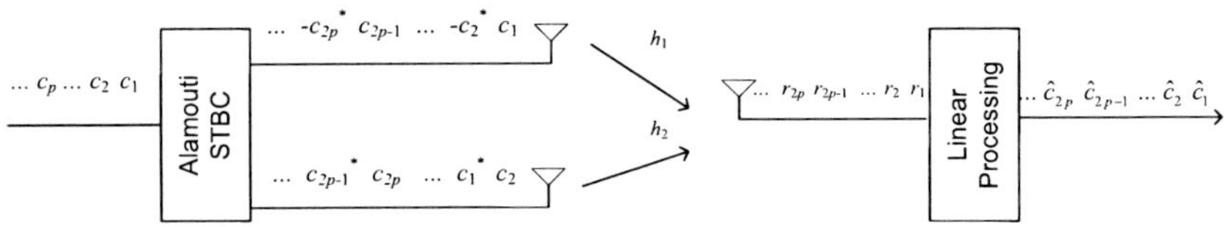

Fig. 1.3 Transmit diversity with two transmit and one receive antenna

In this case, h_1 and h_2 are the CSI from the first and second transmit antenna to the single receive antenna respectively, and they are assumed to remain unchanged for two symbol periods. The received signals at time $2p$-1 and $2p$ can be expressed respectively as:

$$r_{2p-1} = \frac{1}{\sqrt{2}}\left(h_1 c_{2p-1} + h_2 c_{2p}\right) + \eta_{2p-1},$$

$$r_{2p} = \frac{1}{\sqrt{2}}\left(h_2 c_{2p-1}^* - h_1 c_{2p}^*\right) + \eta_{2p}, \tag{1.4}$$

where the factor $1/\sqrt{2}$ accounts for the power normalization, η_{2p-1} and η_{2p} are AWGN at the receiver at time $2p$-1 and $2p$ respectively.

Assuming that perfect CSI is known to the receiver, the transmitted data symbols can be recovered by linear combining as shown below:

$$\hat{c}_{2p-1} = h_1^* r_{2p-1} + h_2 r_{2p}^* = \frac{1}{\sqrt{2}}(|h_1|^2 + |h_2|^2)c_{2p-1} + h_1^* \eta_{2p-1} + h_2 \eta_{2p}^*,$$

$$\hat{c}_{2p} = h_2^* r_{2p-1} - h_1 r_{2p}^* = \frac{1}{\sqrt{2}}(|h_1|^2 + |h_2|^2)c_{2p} + h_2^* \eta_{2p-1} - h_1 \eta_{2p}^*. \tag{1.5}$$

Each of the resultant signals in (1.5) are similar to those in (1.3). Therefore, the diversity order of the above O-STBC system with simple linear receiver processing is the same as the corresponding receive diversity with MRC. This makes O-STBC very attractive. A detailed study of transmit diversity based on STBC will be presented in Chapter 2, while we continue to introduce the signal models for MIMO channel and STBC transmission in this chapter.

1.3 Notations and Abbreviations

Major notations employed in this monograph are:

$*$	Hadamard product
$\otimes$	Kronecker product
$\mathbf{0}_n$	zero matrix of dimension n-by-n
A, a	scalar
$\mathbf{a}$	column vector
$\mathbf{A}$	matrix
$\mathbf{I}_n$	identity matrix of dimension n-by-n
j	square root of -1
$(\,.\,)^*$	complex conjugate
$(\,.\,)^H$	complex conjugate transpose / hermitian
$(\,.\,)^I$	imaginary part of a complex element, vector or matrix
$(\,.\,)^R$	real part of a complex element, vector or matrix
$(\,.\,)^T$	transpose
$|\,.\,|$	magnitude of complex element
$\|\,.\,\|^2$	Frobenius norm
$\lceil x \rceil$	smallest integer larger than x
$\mathrm{Det}(\mathbf{M})$	determinant of a matrix $\mathbf{M}$
$\mathrm{E}(.)$	expectation operator
$\max(.)$	maximization operator
$\min(.)$	minimization operator
$\mathrm{Rank}(\mathbf{M})$	rank of a matrix $\mathbf{M}$
$\mathrm{Tr}(\mathbf{M})$	trace of a matrix $\mathbf{M}$

Major abbreviations are:

AOD	amicable orthogonal design
AWGN	additive white Gaussian noise
BER	bit error rate
BLER	block error rate
bps/Hz	bits per sec per hertz
CSI	channel state information
CR	constellation rotation
DSTM	differential space-time modulation
GCLT	group-constrained linear transformation
JD	joint detection
MDC-QOC	minimum–decoding–complexity quasi-orthogonality constraints
MDC-QOSTBC	minimum–decoding–complexity quasi-orthogonal STBC
MIMO	multiple input multiple output
ML	maximum-likelihood
MSD	maximal symbol-wise diversity
OD	orthogonal design
O-STBC	orthogonal space-time block code
PSK	Phase shift keying
QAM	quadrature amplitude modulation
QOC	quasi-orthogonality constraints
QO-STBC	quasi-orthogonal space-time block code
SD	sphere decoding
SNR	signal-to-noise ratio
STBC	space-time block code

1.4 Signal Model of MIMO Channel and STBC

1.4.1 Signal model of MIMO channel

We consider a general MIMO wireless communication system with N_T transmit antennas at the base station and N_R receive antennas at the mobile, as shown in Fig. 1.4. At each time slot p, the signal $x_p^{(i)}$ is transmitted from the ith transmit antennas, where $i = 1, 2, ..., N_T$. The channel is assumed to be a flat fading channel and the path gain from transmit antenna i to receive antenna k is denoted as $h_{i,k}$. It is assumed that $h_{i,k}$ and $h_{l,q}$ are independent for $1 \leq i, l \leq N_T, 1 \leq k, q \leq N_R$, and for different i, k and l, q pairs. This condition is satisfied if the antennas are well separated by more than half of the wavelength of the transmitted wave, or by using antennas with different polarization. Since we shall focus on transmit diversity with just one receive antenna in our study, the receive antenna index k will later be omitted when not used, i.e. $h_{i,k}$ will just be written as h_i for simplicity. The flat fading path gains are modeled as independent complex Gaussian random variables with variance 0.5 per real dimension, i.e. $h_{i,k} = \alpha_{i,k} \exp(j\theta_{i,k})$, where $\alpha_{i,k}$ follows the Rayleigh distribution and $\theta_{i,k}$ is uniformly distributed. The channel fading is assumed to be quasi-static, i.e. the path gains are assumed to be constant over a frame of length F and only vary from frame to frame.

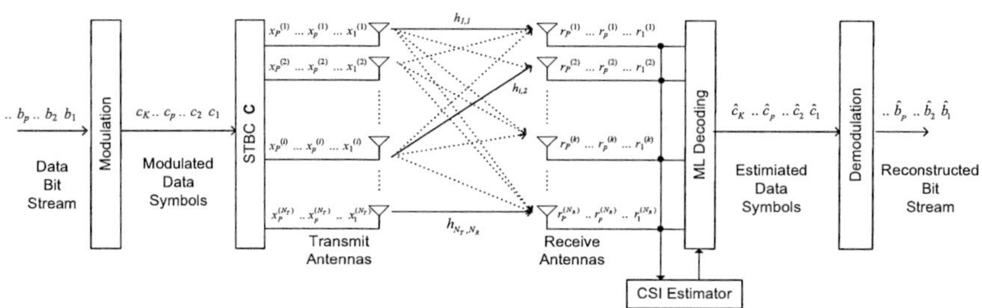

Fig. 1.4 MIMO wireless communication model

We assume that the transmitted signal $x_p^{(i)}$ has unit power and E_s is the total energy transmitted from all antennas. Therefore, the energy transmitted from each transmit antenna is E_s / N_T. Signals arriving at different receive antennas undergo independent fading. The signal at each receive antenna is a noisy superposition of different faded versions of the N_T transmitted signals. At time p, the signal $r_p^{(k)}$ received at the kth antenna, where $k = 1, 2, \ldots, N_R$, is given by

$$r_p^{(k)} = \sqrt{\frac{E_s}{N_T}} \sum_{i=1}^{N_T} \left(h_{i,k} x_p^{(i)} + \eta_p^{(k)} \right), \tag{1.6}$$

where the noise samples $\eta_p^{(k)}$ are independent samples of a zero-mean complex white Gaussian random variable with variance $N_o / 2$ per real dimension. Since the total energy of the symbols transmitted from all transmit antennas is normalized to be E_s, the average energy of the received signal at each receive antenna is E_s and the SNR ρ is E_s / N_o per receive antenna. Without loss of generality, we may assume E_s to be equal to one unless it is stated otherwise.

The signals (1.6) collected by all the receive antennas may be expressed compactly in matrix form as follows:

$$\begin{bmatrix} r_p^{(1)} \\ r_p^{(2)} \\ \vdots \\ r_p^{(k)} \\ \vdots \\ r_p^{(N_R)} \end{bmatrix} = \sqrt{\frac{E_s}{N_T}} \begin{bmatrix} h_{1,1} & h_{2,1} & \cdots & h_{i,1} & \cdots & h_{N_T,1} \\ h_{1,2} & h_{2,2} & \cdots & h_{i,2} & \cdots & h_{N_T,2} \\ \vdots & \vdots & \ddots & \vdots & \ddots & \vdots \\ h_{1,k} & h_{2,k} & \cdots & h_{i,k} & \cdots & h_{N_T,k} \\ \vdots & \vdots & \ddots & \vdots & \ddots & \vdots \\ h_{1,N_R} & h_{2,N_R} & \cdots & h_{i,N_R} & \cdots & h_{N_T,N_R} \end{bmatrix} \begin{bmatrix} x_p^{(1)} \\ x_p^{(2)} \\ \vdots \\ x_p^{(i)} \\ \vdots \\ x_p^{(N_T)} \end{bmatrix} + \begin{bmatrix} \eta_p^{(1)} \\ \eta_p^{(2)} \\ \vdots \\ \eta_p^{(k)} \\ \vdots \\ \eta_p^{(N_R)} \end{bmatrix}, \tag{1.7}$$

where the N_R–by–N_T MIMO channel $\mathcal{H}$ is defined as:

$$\mathcal{H} = \begin{bmatrix} h_{1,1} & h_{2,1} & \cdots & h_{i,1} & \cdots & h_{N_T,1} \\ h_{1,2} & h_{2,2} & \cdots & h_{i,2} & \cdots & h_{N_T,2} \\ \vdots & \vdots & \ddots & \vdots & \ddots & \vdots \\ h_{1,k} & h_{2,k} & \cdots & h_{i,k} & \cdots & h_{N_T,k} \\ \vdots & \vdots & \ddots & \vdots & \ddots & \vdots \\ h_{1,N_R} & h_{2,N_R} & \cdots & h_{i,N_R} & \cdots & h_{N_T,N_R} \end{bmatrix}. \tag{1.8}$$

The information capacity achieved by such an open-loop MIMO channel is [1,2]:

$$C_{\text{channel}} = \mathrm{E}_{\mathcal{H}}\left\{\log\det\left(\mathbf{I}_{N_R} + \frac{\rho}{N_T}\mathcal{H}\mathcal{H}^{\mathrm{H}}\right)\right\} \quad \text{bps/channel use}, \quad (1.9)$$

where $\mathrm{E}\{.\}$ represents the expectation operation and bps stands for bits per sec.

1.4.2 Signal model of STBC

Suppose that a generic STBC codeword is transmitted from N_T transmit antennas to N_R receive antennas over an interval of P symbol periods. The propagation channel condition is time-invariant within a frame length of F symbol periods ($F \geq P$) and is known to the receiver. The transmitted codeword can be written as a $P \times N_T$ matrix $\mathbf{C}$ that contains K complex constellation symbols. Its code length is P, and its code rate is defined as $R = K / P$. Following the model in [11], $\mathbf{C}$ can be expressed as:

$$\mathbf{C} = \sum_{i=1}^{K}\left(c_i^{\mathrm{R}}\mathbf{A}_i + jc_i^{\mathrm{I}}\mathbf{B}_i\right), \quad (1.10)$$

where the information symbols are $c_i = c_i^{\mathrm{R}} + jc_i^{\mathrm{I}}$, and c_i^{R} and c_i^{I} are the real (I) and the imaginary (Q) components of c_i. Matrices $\mathbf{A}_i$ and $\mathbf{B}_i$, both of dimension $P \times N_T$, are called the *"dispersion matrices"* of the STBC.

To illustrate all the above definitions, we again use the Alamouti STBC in Fig. 1.3 as an example. Using the signal model in (1.10), we can write the Alamouti STBC codeword as follows:

$$\mathbf{C}_{\text{Alamouti}} = \begin{bmatrix} c_1 & c_2 \\ -c_2^{*} & c_1^{*} \end{bmatrix}, \quad (1.11)$$

where the corresponding dispersion matrices are:

$$\mathbf{A}_1 = \begin{bmatrix} 1 & 0 \\ 0 & 1 \end{bmatrix}, \mathbf{A}_2 = \begin{bmatrix} 0 & 1 \\ -1 & 0 \end{bmatrix}, \mathbf{B}_1 = \begin{bmatrix} 1 & 0 \\ 0 & -1 \end{bmatrix}, \mathbf{B}_2 = \begin{bmatrix} 0 & 1 \\ 1 & 0 \end{bmatrix}. \quad (1.12)$$

From the above example, it should be clear that the matrix $\mathbf{A}_i$ "disperses" the real part of the symbol c_i on different antenna and time positions, while the matrix $\mathbf{B}_i$ does the same for the imaginary part of c_i, hence the name "dispersion matrix" for them.

The row of the STBC codeword represents the signal to be transmitted at a particular time slot, while the column of the STBC codeword represents the signal to be transmitted at a particular transmit antennas. One can also easily see that for $N_\mathrm{T} = 2$ transmit antennas, the Alamouti STBC takes $P = 2$ symbols period to transmit $K = 2$ complex symbols, hence it has a code rate R of $K/P = 1$.

For a given number of transmit antennas, the design of a STBC depends crucially on the code parameters P, K, and the dispersion matrices $\{\mathbf{A}_i, \mathbf{B}_i\}$. With the representation of STBC in (1.10), the transmitted and received signals are related by [11]:

$$\tilde{\mathbf{r}} = \sqrt{E_\mathrm{s}/N_\mathrm{T}}\,\mathbf{H}\tilde{\mathbf{c}} + \tilde{\boldsymbol{\eta}}, \tag{1.13}$$

where

$$\tilde{\mathbf{r}} = \begin{bmatrix} \mathbf{r}_1^R \\ \mathbf{r}_1^I \\ \vdots \\ \mathbf{r}_{N_R}^R \\ \mathbf{r}_{N_R}^I \end{bmatrix}, \tilde{\mathbf{c}} = \begin{bmatrix} c_1^R \\ c_1^I \\ \vdots \\ c_K^R \\ c_K^I \end{bmatrix}, \tilde{\boldsymbol{\eta}} = \begin{bmatrix} \boldsymbol{\eta}_1^R \\ \boldsymbol{\eta}_1^I \\ \vdots \\ \boldsymbol{\eta}_{N_R}^R \\ \boldsymbol{\eta}_{N_R}^I \end{bmatrix},$$

$$\mathbf{H} = \begin{bmatrix} \mathcal{A}_1\tilde{\mathbf{h}}_1 & \mathcal{B}_1\tilde{\mathbf{h}}_1 & \cdots & \mathcal{A}_K\tilde{\mathbf{h}}_1 & \mathcal{B}_K\tilde{\mathbf{h}}_1 \\ \vdots & \vdots & \ddots & \vdots & \vdots \\ \mathcal{A}_1\tilde{\mathbf{h}}_{N_R} & \mathcal{B}_1\tilde{\mathbf{h}}_{N_R} & \cdots & \mathcal{A}_K\tilde{\mathbf{h}}_{N_R} & \mathcal{B}_K\tilde{\mathbf{h}}_{N_R} \end{bmatrix},$$

$$\mathcal{A}_q = \begin{bmatrix} \mathbf{A}_q^R & -\mathbf{A}_q^I \\ \mathbf{A}_q^I & \mathbf{A}_q^R \end{bmatrix}, \mathcal{B}_q = \begin{bmatrix} -\mathbf{B}_q^I & -\mathbf{B}_q^R \\ \mathbf{B}_q^R & -\mathbf{B}_q^I \end{bmatrix}, \tilde{\mathbf{h}}_i = \begin{bmatrix} \mathbf{h}_i^R \\ \mathbf{h}_i^I \end{bmatrix}.$$

In the above equation, $\mathbf{r}_i$ and $\boldsymbol{\eta}_i$ ($1 \leq i \leq N_R$) are $P \times 1$ column vectors which contain the received signals and AWGN noises for the ith receive antenna respectively, over P symbol periods. $\mathbf{H}$ of dimension $2PN_R \times 2K$ is called the *equivalent channel matrix*, $\mathbf{h}_i$ is a $N_T \times 1$ column vector that contains the fading coefficients of the spatial sub-channels between the N_T transmit antennas and i^th receive antenna. The normalization factor

$\sqrt{E_s/N_T}$ in (1.13) ensures that the SNR $\rho = E_s/N_o$ is the same at each receive antenna, regardless of what N_T is.

Using the Alamouti STBC in (1.12) as an example, its equivalent channel matrix $\mathbf{H}$ can be computed using the following four matrices:

$$\mathcal{A}_1 = \begin{bmatrix} 1 & 0 & 0 & 0 \\ 0 & 1 & 0 & 0 \\ 0 & 0 & 1 & 0 \\ 0 & 0 & 0 & 1 \end{bmatrix}, \mathcal{A}_2 = \begin{bmatrix} 0 & 1 & 0 & 0 \\ -1 & 0 & 0 & 0 \\ 0 & 0 & 0 & 1 \\ 0 & 0 & -1 & 0 \end{bmatrix},$$

$$\mathcal{B}_1 = \begin{bmatrix} 0 & 0 & -1 & 0 \\ 0 & 0 & 0 & 1 \\ 1 & 0 & 0 & 0 \\ 0 & -1 & 0 & 0 \end{bmatrix}, \mathcal{B}_2 = \begin{bmatrix} 0 & 0 & 0 & -1 \\ 0 & 0 & -1 & 0 \\ 0 & 1 & 0 & 0 \\ 1 & 0 & 0 & 0 \end{bmatrix}. \tag{1.14}$$

Basic design requirements on the dispersion matrices include the three *Power Distribution Constraints* [11]:

$$\text{(i)} \quad \sum_{i=1}^{K} \left[\text{Tr}\left(\mathbf{A}_i^H \mathbf{A}_i \right) + \text{Tr}\left(\mathbf{B}_i^H \mathbf{B}_i \right) \right] = 2PN_T;$$

$$\text{(ii)} \quad \text{Tr}\left(\mathbf{A}_i^H \mathbf{A}_i \right) = \text{Tr}\left(\mathbf{B}_i^H \mathbf{B}_i \right) = \frac{PN_T}{K}, \qquad 1 \le i \le K; \tag{1.15}$$

$$\text{(iii)} \quad \mathbf{A}_i^H \mathbf{A}_i = \mathbf{B}_i^H \mathbf{B}_i = \frac{P}{K}\mathbf{I}_{N_T}, \qquad 1 \le i \le K.$$

And Tr(.) denotes the trace, or sum of all diagonal elements, of a matrix.

The constraints (i)–(iii) in (1.15) govern the distribution of transmission power across space and time in the following ways: Constraint (i) ensures that the total average transmitted power is normalized to PN_T, i.e. $\text{E}[\text{Tr}(\mathbf{CC}^H)] = PN_T$. Constraint (ii) is more restrictive and ensures that each of the I and Q signals (c_i^R and c_i^I) is transmitted with the same overall power over P symbol durations from all antennas. Constraint (iii) is the most stringent: it forces the symbols c_i^R and c_i^I to have equal energy in all spatial directions during *each* symbol duration. It is pointed out in [11] that codes satisfying the more stringent constraint in (1.15) (iii) generally give lower error rates in a feedforward or blind MIMO scheme. This agrees with the idea of [12] which states that since the encoder of a feedforward or blind MIMO system does not have any information about the channel, it should transmit the code symbols uniformly in all space and time directions, assuming that all spatial channels are independent and identically distributed.

It is further proposed in [13] that constraint (1.15) (iii) is the *maximal symbol-wise diversity* (MSD) condition, which guarantees full diversity protection for one-symbol error events that have the smallest Euclidean distance. As suggested in [13], intuitively it is obvious that if a code cannot provide full diversity protection against the one-symbol error event, it cannot provide full diversity protection aginst all error events.

As mentioned earlier, besides the dispersion matrices $\mathbf{A}$ and $\mathbf{B}$, the design of STBC also depends on the number of information symbols transmitted within a codeword K, and the code length P. These two values are generally limited by the following constraints in (1.16) and (1.17):

$$\text{transmit diversity level} \leq \min(N_T, P), \tag{1.16}$$

$$K \leq P\min(N_T, N_R). \tag{1.17}$$

Constraint (1.16) arises because the diversity order of a STBC is determined by the rank of its codeword (this will be elaborated in Section 1.5). Since the STBC codeword has dimension $P \times N_T$, its rank is limited by the minimum value of N_T and P.

To understand constraint (1.17), let us refer to the signal model in (1.13). The K transmitted data symbols is represented by $2K$ unknown real values in $\tilde{\mathbf{c}}$. In order to solve for $\tilde{\mathbf{c}}$ at the receiver without ambiguity, there must be at least $2K$ linearly independent equations in (1.13). This implies that the equivalent channel matrix $\mathbf{H}$, which has dimension $2PN_R \times 2K$, must have $PN_R \geq K$. In addition, the dispersion matrices $\mathbf{A}$ and $\mathbf{B}$, each of size $P \times N_T$, must have K independent basis, this requires $PN_T \geq K$. Combining this and the earlier $PN_R \geq K$ condition gives (1.17).

In order to achieve *full transmit diversity* of order N_T from the N_T transmit antennas, (1.16) requires that $P \geq N_T$. Hence a square code design with $P = N_T$ gives the minimum possible code length for achieving full diversity, and STBC with such property is commonly called STBC with *minimum delay*. Furthermore, if only one receive antenna is used, i.e. $N_R = 1$, we get $K \leq P$ from (1.17). Hence the maximum achievable code rate $R = K / P$ of a transmit diversity scheme with only one receive antenna is one.

The capacity, C_{STBC}, achievable by a STBC is [11]:

$$C_{STBC} = \frac{1}{2P} E_H \left[\log \det \left(I_{2N_R P} + \frac{\rho}{N_T} HH^T \right) \right] \quad \text{bps/channel use. (1.18)}$$

This is different from (1.9) as (1.9) is the capacity formula for a MIMO channel, while (1.18) is the capacity achievable by a STBC.

1.5 Design Criteria and Performance Measure of STBC

Guey *et al.* pointed out in [14] that the critical parameter for evaluating the performance of a space-time code in slow flat fading channel is the rank of the codeword difference matrix. In [4], Tarokh *et al.* further showed that the minimum rank of the codeword difference matrix quantifies the *diversity gain*, while the minimum product of the non-zero eigenvalues of the codeword distance matrix with the minimum rank quantifies the *coding gain*, of the space-time code. They will be reviewed in the following.

We assume that all the codewords have equal transmission probability and let P(**C**→**E**) denote the probability that the codeword **C** is transmitted but the receiver decides erroneously in favor of another codeword **E**. This probability term is commonly called the *pair-wise error probability* (PEP). With ideal CSI, PEP is well approximated as follows [4]:

$$P\left(C \rightarrow E \mid h_{i,k}, i = 1,2,...,N_T; k = 1,2,...,N_R \right) \leq \exp\left(-d^2 \left(C,E \right) \frac{E_s}{4N_o} \right), (1.19)$$

where

$$d^2(\mathbf{C},\mathbf{E}) = \sum_{k=1}^{N_R} \sum_{p=1}^{P} \left| \sum_{i=1}^{N_T} h_{i,k}(c_{p,i} - e_{p,i}) \right|^2 ,$$

$c_{p,i}$ and $e_{p,i}$ are the entries in the pth row and i^{th} column of $\mathbf{C}$ and $\mathbf{E}$.

To simplify the PEP expression, we first define two matrices, the first is the *codeword difference matrix* $\mathbf{B}_{\mathbf{CE}}$ of size $P \times N_T$, which is defined as [4]:

$$\mathbf{B}_{\mathbf{CE}} = \mathbf{C} - \mathbf{E} , \tag{1.20}$$

and the second is the *codeword distance matrix* $\mathbf{A}_{\mathbf{CE}}$ of size $N_T \times N_T$, which is defined as [15]:

$$\mathbf{A}_{\mathbf{CE}} = \mathbf{B}_{\mathbf{CE}}^{\mathrm{H}} \mathbf{B}_{\mathbf{CE}} . \tag{1.21}$$

Furthermore, we also define $\lambda_1, \lambda_2, \ldots, \lambda_D$ as the non-zero eigenvalues of $\mathbf{A}_{\mathbf{CE}}$, where D denotes the rank of $\mathbf{A}_{\mathbf{CE}}$, which is the same as the rank of $\mathbf{B}_{\mathbf{CE}}$. At high SNR, the upper bound of the PEP in (1.19) can be simplified as [4]:

$$P(\mathbf{C} \to \mathbf{E}) \le \left[\left(\prod_{i=1}^{D} \lambda_i \right) \left(\frac{E_s}{4N_o} \right)^{-D} \right]^{-N_R} . \tag{1.22}$$

From (1.22), we can define the following two quantities to account for the decoding performance of a STBC:

- *Transmit Diversity Gain*: the minimum rank D of the matrix $\mathbf{A}_{\mathbf{CE}}$ over all pairs of distinct codewords. It accounts for the slope of the bit error rate (BER) or block error rate (BLER) curve of the STBC. A STBC is said to achieve *full diversity* if $D = N_T$.
- *Diversity Product*: defined as follows in [16] for a full diversity code,

$$\varsigma = \frac{1}{2\sqrt{N_T}} \min_{\forall \mathbf{C} \ne \mathbf{E}} \left| \mathrm{Det}(\mathbf{A}_{\mathbf{CE}}) \right|^{1/(2P)}$$

$$= \frac{1}{2\sqrt{N_T}} \min_{\forall \mathbf{C} \ne \mathbf{E}} \left(\prod_{i=1}^{D} \lambda_i \right)^{1/(2P)} , \tag{1.23}$$

where the factor $1/\left(2\sqrt{N_\mathrm{T}}\right)$ guarantees that $0 \leq \zeta \leq 1$, and Det(.) denotes the determinant of a matrix. Diversity product accounts for the *coding gain*, which determines the left or right shift of the BER or BLER curve of the STBC.

Based on the above, a set of code design criteria for space-time codes, commonly called the *Rank & Determinant Criteria*, can be stated as follows [4]:

- *Rank Criterion*: To maximize the diversity gain of the STBC, maximize the minimum rank D of the matrix $\mathbf{A_{CE}}$ over all distinct pairs of codewords. Hence, in order to achieve the maximum diversity order, the matrix $\mathbf{A_{CE}}$ has to be full rank, i.e. $D = N_\mathrm{T}$, for any codeword pair $\mathbf{C}$ and $\mathbf{E}$.
- *Determinant Criterion*: To maximize the coding gain of the STBC, maximize the minimum product of non-zero eigenvalues of the matrix $\mathbf{A_{CE}}$ which has the minimum rank. When a code achieves full diversity, this determinant criterion implies the maximization of the diversity product in (1.23).

Based on the channel and STBC models presented in this chapter, we will give a detailed overview on the design of O-STBC and QO-STBC in the next chapter.

Chapter 2

Orthogonal and Quasi-Orthogonal Space-Time Block Code

In Chapter 1, we discussed the advantages of using multiple transmit antennas to provide transmit diversity for wireless communication. In particular, we have shown that the Alamouti STBC is able to provide transmit diversity that achieves the same diversity gain as receive diversity by using simple linear receiver processing. The adoption of Almouti STBC in the 3GPP standards has sparked numerous research works to design similar schemes that is able to provide higher-order transmit diversity with equally low decoding complexity. These research efforts have led to the discoveries of the *Orthogonal Space-Time Block Code* (O-STBC) and *Quasi-Orthogonal Space-Time Block Code* (QO-STBC).

2.1 Orthogonal Space-Time Block Code

2.1.1 Benefits of O-STBC

The discovery of Alamouti STBC, a simple and yet effective transmit diversity for two transmit antennas, has spurred the idea that similar schemes that support a larger number of transmit antennas with higher diversity order may be possible. Hence in [6], a group of researchers from AT&T generalize Alamouti STBC to more than two transmit antennas by constructing the *Orthogonal Space-Time Block Code* (O-STBC), a class of STBC that achieves:

20

- Full transmit diversity, i.e. using multiple transmit antennas to achieve the same number of diversity gain as receive MRC diversity with the same number of receive antennas.
- ML decoding by simple linear processing, i.e. separate detection of the real and imaginary parts of individual QAM symbols, or symbol–by–symbol detection of individual PSK symbols.

Besides the above two properties, it was later found in [17] and [18] that O-STBC also maximizes the received SNR and minimizes the PEP.

Mathematically, an O-STBC codeword $\mathbf{C}$ is defined to be one that satisfies [6]:

$$\mathbf{C}^H\mathbf{C} = \left(\sum_{i=1}^{K} \left(\left|c_i^R\right|^2 + \left|c_i^I\right|^2 \right) \right) \mathbf{I}_{N_T}. \tag{2.1}$$

It has been shown in [17,18] that to satisfy (2.1), the dispersion matrices $\mathbf{A}_i$ and $\mathbf{B}_i$ of an O-STBC must satisfy the following constraints:

$$\text{(i)} \quad \begin{aligned} \mathbf{A}_i^H\mathbf{A}_i &= \mathbf{I}_{N_T}, \\ \mathbf{B}_i^H\mathbf{B}_i &= \mathbf{I}_{N_T}, \end{aligned} \qquad 1 \le i \le K;$$

$$\text{(ii)} \quad \begin{aligned} \mathbf{A}_i^H\mathbf{A}_k &= -\mathbf{A}_k^H\mathbf{A}_i, \\ \mathbf{B}_i^H\mathbf{B}_k &= -\mathbf{B}_k^H\mathbf{B}_i, \end{aligned} \qquad 1 \le i \ne k \le K; \tag{2.2}$$

$$\text{(iii)} \quad \mathbf{A}_i^H\mathbf{B}_k = \mathbf{B}_k^H\mathbf{A}_i, \qquad 1 \le i,k \le K.$$

In addition, it is also desired that the O-STBC has a high code rate and low decoding delay, i.e. K as large as possible and P as small as possible. Based on different mathematical approaches, square O-STBC with minimum decoding delay ($P = N_T$ [1]) has been designed in [17] and [19] by using *Amicable Orthogonal Design* (AOD) and *Clifford Algebra* respectively.

---

[1] From (1.16), the diversity gain provided by a STBC is bounded by the smaller value of the number of transmit antennas N_T and the code length P. To achieve full transmit diversity of N_T, P must greater or equal to N_T. Hence, the minimum code length P (corresponding to minimum decoding delay) is N_T.

2.1.2 *Background of amicable orthogonal design*

To distinguish between AOD matrices and STBC dispersion matrices, in this section the mathematical symbols related to an AOD will be underlined, while those related to a STBC will not.

<u>Definition 2.1</u> [20]: A weighting matrix $\underline{\mathbf{W}}$ of weight w and order n is an $n \times n$ matrix with entries from $\{0, +1, -1\}$ satisfying

$$\underline{\mathbf{W}}^{\mathrm{T}}\underline{\mathbf{W}} = w\mathbf{I}_n. \tag{2.3}$$

We shall denote such a matrix by $\underline{\mathbf{W}}(n, w)$.

<u>*Definition 2.2*</u> [20]: An *Orthogonal Design* (OD) matrix $\underline{\mathbf{X}}$ of order n and of type $(u_1, \ldots, u_s)$, u_i positive integers, is an $n \times n$ matrix with entries from $\{0, \pm \underline{x}_1, \ldots, \pm \underline{x}_s\}$ satisfying

$$\underline{\mathbf{X}}^{\mathrm{T}}\underline{\mathbf{X}} = \left(\sum_{i=1}^{s} u_i \underline{x}_i^2\right)\mathbf{I}_n. \tag{2.4}$$

$\underline{\mathbf{X}}$ can be expressed as

$$\underline{\mathbf{X}} = \underline{\mathbf{A}}_1\underline{x}_1 + \underline{\mathbf{A}}_2\underline{x}_2 + \cdots + \underline{\mathbf{A}}_s\underline{x}_s, \tag{2.5}$$

where each $\underline{\mathbf{A}}_i$ is $\underline{\mathbf{W}}(n, u_i)$ satisfying:

$$\begin{aligned}
(0) \quad & \underline{\mathbf{A}}_i * \underline{\mathbf{A}}_k = \mathbf{0}, & 1 \le i \ne k \le s; \\
(i) \quad & \underline{\mathbf{A}}_i^{\mathrm{T}}\underline{\mathbf{A}}_i = u_i\mathbf{I}_n, & 1 \le i \le s; \\
(ii) \quad & \underline{\mathbf{A}}_i^{\mathrm{T}}\underline{\mathbf{A}}_k + \underline{\mathbf{A}}_k^{\mathrm{T}}\underline{\mathbf{A}}_i = \mathbf{0}, & 1 \le i \ne k \le s.
\end{aligned} \tag{2.6}$$

And $*$ represents the Hadamard product.

<u>*Lemma 2.1*</u> [21]: The number of variables, s, in an orthogonal design of order $n = 2^a b$, b odd and $a = 4c + d$ where $0 \le d < 4$, is upper bounded as follows:

$$s \le \rho(n), \tag{2.7}$$

where

$$\rho(n) = 8c + 2^d. \tag{2.8}$$

In the special cases of $n = 1, 2, 4$ and 8, $\rho(n) = n$.

Definition 2.3 [22]: Let $\underline{\mathbf{X}}$ and $\underline{\mathbf{Y}}$ be OD of the same order n where $\underline{\mathbf{X}}$ is of type $(u_1, \ldots, u_s)$ on the variables $\{\underline{x}_1, \ldots, \underline{x}_s\}$ and $\underline{\mathbf{Y}}$ is of type $(v_1, \ldots, v_t)$ on the variables $\{\underline{y}_1, \ldots, \underline{y}_t\}$. Similar to $\underline{\mathbf{X}}$ in (2.5), $\underline{\mathbf{Y}}$ can be expressed as

$$\underline{\mathbf{Y}} = \underline{\mathbf{B}}_1\underline{y}_1 + \underline{\mathbf{B}}_2\underline{y}_2 + \cdots + \underline{\mathbf{B}}_t\underline{y}_t, \tag{2.9}$$

$\underline{\mathbf{X}}$ and $\underline{\mathbf{Y}}$ are said to be *Amicable Orthogonal Design* (AOD) if

$$\underline{\mathbf{X}}^{\mathrm{T}}\underline{\mathbf{Y}} = \underline{\mathbf{Y}}^{\mathrm{T}}\underline{\mathbf{X}}. \tag{2.10}$$

A necessary and sufficient condition for the AOD as defined in *Definition 2.3* to exist is that there must exist a family of matrices $\{\underline{\mathbf{A}}_1, \ldots, \underline{\mathbf{A}}_s ; \underline{\mathbf{B}}_1, \ldots, \underline{\mathbf{B}}_t\}$ of order n with entries from $\{0, 1, -1\}$ satisfying:

$$
\begin{aligned}
(0) \quad & \underline{\mathbf{A}}_i * \underline{\mathbf{A}}_p = \mathbf{0}, & & 1 \leq i \neq p \leq s; \\
& \underline{\mathbf{B}}_k * \underline{\mathbf{B}}_q = \mathbf{0}, & & 1 \leq k \neq q \leq t; \\
(i) \quad & \underline{\mathbf{A}}_i^{\mathrm{T}}\underline{\mathbf{A}}_i = u_i\mathbf{I}_n, & & 1 \leq i \leq s; \\
& \underline{\mathbf{B}}_k^{\mathrm{T}}\underline{\mathbf{B}}_k = v_k\mathbf{I}_n, & & 1 \leq k \leq t; \\
(ii) \quad & \underline{\mathbf{A}}_i^{\mathrm{T}}\underline{\mathbf{A}}_p + \underline{\mathbf{A}}_p^{\mathrm{T}}\underline{\mathbf{A}}_i = \mathbf{0}, & & 1 \leq i \neq p \leq s; \\
& \underline{\mathbf{B}}_k^{\mathrm{T}}\underline{\mathbf{B}}_q + \underline{\mathbf{B}}_q^{\mathrm{T}}\underline{\mathbf{B}}_k = \mathbf{0}, & & 1 \leq k \neq q \leq t; \\
(iii) \quad & \underline{\mathbf{A}}_i^{\mathrm{T}}\underline{\mathbf{B}}_k - \underline{\mathbf{B}}_k^{\mathrm{T}}\underline{\mathbf{A}}_i = \mathbf{0}, & & 1 \leq i \leq s, 1 \leq k \leq t.
\end{aligned}
\tag{2.11}
$$

Hence (2.11) (0)-(iii) serve as the criteria to construct AOD matrices.

In *Definition 2.3*, the number of variables in the OD $\underline{\mathbf{X}}$ of an AOD is denoted by s, while the number of variables in the OD $\underline{\mathbf{Y}}$ of an AOD is denoted by t. The upper bounds of s and t are summarized in *Lemma 2.2*.

Lemma 2.2 [21]: For an AOD of order n, where $n=2^a b$, a and b are both integers and b is odd and $a = 4c + d$ where $0 \leq d < 4$,

$$
\begin{aligned}
t &\leq \rho(n), \\
s &\leq \rho_t(n),
\end{aligned}
\tag{2.12}
$$

where $\rho(n)$ is as defined in *Lemma 2.1*, and $\rho_t(n)$ is defined as follows:

$$\rho_t(n) = 8c - t + \delta + 1, \tag{2.13}$$

where the values of δ depend on the values of t and d, and are given in Table 2.1 [21].

Table 2.1 Values of δ for an AOD

$t(\mathrm{mod}4)$ \ d	0	1	2	3
0	0	1	3	7
1	1	2	3	5
2	-1	3	4	5
3	-1	1	5	6

For $n = 4$ and 8, the corresponding values of s and t are listed in Table 2.2 and Table 2.3 respectively. For example, for $n = 4$, we know from (2.12) and (2.8) that:

- t must be less than or equal to $\rho(4) = 4$, so t may take values from 0 to 4, as shown in Table 2.2.
- s must be less than or equal to $\rho_t(4)$ according to (2.13). Since t has a range of values as explained above, s will have different upper bound values. For example, in Table 2.2 for $t = 3$, $s \leq \rho_3(4) = 3$.

From Table 2.2 and Table 2.3, we note that $\rho_t(n)$ is a decreasing function of t.

Table 2.2 Values of s and t for $n = 4$

t, number of variables in $\mathbf{Y}$	s, number of variables in $\mathbf{X}$	Max($s+t$), max. number of variables in AOD
4	$\leq \rho_4(4) = 0$	4
3	$\leq \rho_3(4) = 3$	6
2	$\leq \rho_2(4) = 3$	5
1	$\leq \rho_1(4) = 3$	4
0	$\leq \rho_0(4) = 4$	4

Table 2.3 Values of s and t for $n = 8$

t, number of variables in $\underline{\mathbf{Y}}$	s, number of variables in $\underline{\mathbf{X}}$	Max$(s+t)$, max. number of variables in AOD
8	$\leq \rho_8(8) = 0$	8
7	$\leq \rho_7(8) = 0$	7
6	$\leq \rho_6(8) = 0$	6
5	$\leq \rho_5(8) = 1$	6
4	$\leq \rho_4(8) = 4$	8
3	$\leq \rho_3(8) = 4$	7
2	$\leq \rho_2(8) = 4$	6
1	$\leq \rho_1(8) = 5$	6
0	$\leq \rho_0(8) = 8$	8

Lemma 2.3 [21]: For an AOD of order n, where $n=2^a b$, a and b are both integers and b is odd, the total number of variables in an AOD (i.e. $s + t$) is upper bounded by $2a + 2$, and that bound can be achieved.

2.1.3　Construction of O-STBC and its rate limitation

By comparing (2.11) (i)-(iii) with the O-STBC constraints in (2.2) (i)-(iii), it is clear that an AOD can be used to construct an O-STBC [17]. The order n of an AOD will correspond to the number of transmit antennas N_T, while the code length p of the O-STBC will also be n as AOD is a square design. As a result, the dispersion matrices $\mathbf{A}$ and $\mathbf{B}$ of an O-STBC can be obtained directly from the $\underline{\mathbf{A}}$ and $\underline{\mathbf{B}}$ matrices of an AOD. In addition, the code rate R of an O-STBC will be related to the maximum number of variable of an AOD by:

$$R = \frac{\max(s+t)}{2n},$$
(2.14)

where the factor of 2 in the denominator of (2.14) is due to the fact that the code rate of an O-STBC is defined to be the number of *complex* symbols transmitted over a period of time, while a variable in AOD only represents a *real* symbol, which is half of a complex symbol. For example, when $n=2$ ($a = b = 1$ if $n=2^a b$), according to *Lemma 2.3*, the maximum number of variables, $s+t$, that an AOD can have is $(2a + 2) = 4$, hence by (2.14) the maximum code rate of an O-STBC for two transmit antennas is one, which speaks for the well-known Alamouti STBC.

From Table 2.2 and Table 2.3 (shaded rows), one can easily see that the maximum number of variables in an AOD of order four and eight are 6 and 8 respectively, hence it can be deduced that the maximum achievable code rate of a square O-STBC for four and eight transmit antennas are 3/4 and 1/2 respectively -- this makes the Alamouti STBC the only O-STBC that support full code rate for complex constellation. Two examples of O-STBC for four and eight transmit antennas with rate 3/4 and 1/2 respectively, **C4** from [17] and **C8** from [23], are shown below:

$$\mathbf{C4} = \begin{bmatrix} c_1 & 0 & c_2 & -c_3 \\ 0 & c_1 & c_3^* & c_2^* \\ -c_2^* & -c_3 & c_1^* & 0 \\ c_3^* & -c_2 & 0 & c_1^* \end{bmatrix}, \tag{2.15}$$

$$\mathbf{C8} = \begin{bmatrix} c_1 & c_1 & c_2^* & -c_2^* & c_3^* & -c_3^* & c_4^* & -c_4^* \\ c_1^* & -c_1^* & c_2 & c_2 & c_3 & c_3 & c_4 & c_4 \\ -c_2^* & c_2^* & c_1 & c_1 & c_4 & -c_4 & -c_3 & c_3 \\ -c_2 & -c_2 & c_1^* & -c_1^* & c_4^* & c_4^* & -c_3^* & -c_3^* \\ -c_3^* & c_3^* & -c_4 & c_4 & c_1 & c_1 & c_2 & -c_2 \\ -c_3 & -c_3 & -c_4^* & -c_4^* & c_1^* & -c_1^* & c_2^* & c_2^* \\ -c_4^* & c_4^* & c_3 & -c_3 & -c_2 & c_2 & c_1 & c_1 \\ -c_4 & -c_4 & c_3^* & c_3^* & -c_2^* & -c_2^* & c_1^* & -c_1^* \end{bmatrix}. \tag{2.16}$$

One may notice from *Lemma 2.2* that an AOD only exists for a certain values of n, i.e. $n = 2^a b$, where a and b are both integers and b is odd. In particular, when n is not a power of two, such as $n=3$, AOD does not exist and hence cannot be used to construct O-STBC with an antenna number that is not power of two. Fortunately, it has been shown in [6] that by truncating a column of O-STBC codeword, the resultant codeword is still an O-STBC, which can now be used for less antennas. For example, by removing the last column of **C4** in (2.15), the resultant code **C3** shown in (2.17) remains an O-STBC that supports three transmit antennas with the same code rate of ¾:

$$
\mathbf{C3} = \begin{bmatrix} c_1 & 0 & c_2 \\ 0 & c_1 & c_3^* \\ -c_2^* & -c_3 & c_1^* \\ c_3^* & -c_2 & 0 \end{bmatrix}. \tag{2.17}
$$

Note that **C3** is not a square O-STBC. Several researchers have investigated the maximum achievable code rate of non-square O-STBC [24,25]. They have shown that non-square design leads to O-STBC with higher code rate than square design. Hence [26,27,28] have focused on the construction of non-square O-STBC with higher code rate than square O-STBC. However, as shown in Table 2.4, non-square O-STBC generally require a longer decoding delay due to longer code length.

The maximum achievable code rate of O-STBC by square and non-square designs are summarized in (2.18) and Table 2.4.

$$
R_{\text{square}} \le \frac{\lceil \log_2 N_{\text{T}} \rceil + 1}{2^{\lceil \log_2 N_{\text{T}} \rceil}},
$$

$$
R_{\text{non-square}} \le \frac{\lceil N_{\text{T}}/2 \rceil + 1}{2\lceil N_{\text{T}}/2 \rceil}, \tag{2.18}
$$

where $\lceil x \rceil$ denotes the smallest integer larger than x.

Table 2.4 Maximum achievable code rate of square and non-square O-STBC

No. of tx antennas	O-STBC from square design		O-STBC from non-square design	
	Delay, P	Code Rate, R	Delay, P	Code Rate, R
2	2	1	N. A.	
3, 4	4	3/4		
5, 6	8	1/2	15, 30	4/6
7, 8	8	1/2	56	5/8

2.2 Quasi-Orthogonal Space-Time Block Code

Although O-STBC is attractive for its full transmit diversity and linear ML decoding properties, an O-STBC for more than two transmit antennas cannot achieve full code rate. This implies that transmit diversity based on O-STBC is not as bandwidth efficient as receive diversity, which does not suffer such rate limitation. Hence it is of great interest to design new classes of STBC that can achieve full transmit diversity without the penalty in bandwidth efficiency. This can be achieved by removing the orthogonality constraint in the STBC design, resulting in non-orthogonal STBC, but clearly the low decoding complexity advantage of O-STBC will be lost, which is too high a price to pay.

To achieve full-rate transmission while retaining much of the orthogonality benefits of O-STBC, *Quasi-Orthogonal Space-Time Block Code* (QO-STBC) has been proposed. Three groups of researchers from three major telecommunications laboratories have independently proposed such codes at about the same time. They are Jafarkhani from AT&T Laboratories, Tirkkonen, Boariu and Hottinen from Nokia Research, and Papadias and Foschini from Lucent Technolgies Bell Laboratories. Initially, they gave different names to their codes. For example, Jafarkhani named his code QO-STBC [8], which reflects the property of the code well, and is well adopted in the literature subsequently. On the other hand, the Nokia researchers named their code ABBA [9] to reflect the code structure. The researchers from Lucent Technologies Bell Laboratories did not name their code [10], but gave an insightful explanation that such code could achieve the open-loop capacity of a four-transmit-one-receive antenna system. Bell Laboratories also holds the patent of QO-STBC in the USA and Europe [29,30].

With its quasi-orthogonal code structure, the data symbols in a QO-STBC are separable into groups after matched filtering. The maximum-likelihood (ML) decoding of the QO-STBC can then be performed by jointly detecting the symbols group by group, separately and in parallel. As a result, the decoding complexity of *quasi-orthogonal* STBC is much lower than a general *non-orthogonal* STBC, particularly when the codeword dimension or constellation dimension is big.

Motivated by the advantages of high code rate and low decoding complexity, this monograph is written with the objectives to give a detailed review of current QO-STBC technology, present new QO-STBC designs with minimum or near-minimum decoding complexity, and conduct in-depth studies on their attributes and performance. Both coherent and non-coherent QO-STBC schemes will be discussed.

2.2.1 *Approaching capacity with low decoding complexity*

Early QO-STBCs reported in the literature can be classified into the following three categories.

Rate-1 QO-STBC for four transmit antennas

An example of this code is the QO-STBC proposed by Jafarkhani in [8], which is denoted herein as **J4**:

$$\mathbf{J4} = \begin{bmatrix} \mathbf{A} & \mathbf{B} \\ -\mathbf{B}^* & \mathbf{A}^* \end{bmatrix}$$

$$= \begin{bmatrix} c_1 & c_2 & c_3 & c_4 \\ -c_2^* & c_1^* & -c_4^* & c_3^* \\ -c_3^* & -c_4^* & c_1^* & c_2^* \\ c_4 & -c_3 & -c_2 & c_1 \end{bmatrix}, \tag{2.19}$$

where **A** and **B** are two Alamouti STBCs as follows:

$$\mathbf{A} = \begin{bmatrix} c_1 & c_2 \\ -c_2^* & c_1^* \end{bmatrix}, \ \mathbf{B} = \begin{bmatrix} c_3 & c_4 \\ -c_4^* & c_3^* \end{bmatrix}. \tag{2.20}$$

Another example is the ABBA code proposed by Tirkkonen *et al.* in [9], which is denoted herein as **TBH4**:

$$\mathbf{TBH4} = \begin{bmatrix} \mathbf{A} & \mathbf{B} \\ \mathbf{B} & \mathbf{A} \end{bmatrix}$$

$$= \begin{bmatrix} c_1 & c_2 & c_3 & c_4 \\ -c_2^* & c_1^* & -c_4^* & c_3^* \\ c_3 & c_4 & c_1 & c_2 \\ -c_4^* & c_3^* & -c_2^* & c_1^* \end{bmatrix}. \tag{2.21}$$

Similarly, the $\mathbf{A}$ and $\mathbf{B}$ are two Alamouti STBC defined in (2.20). Due to its codeword structure, it is commonly known in the literature as the ABBA code.

Yet another example is the patented QO-STBC from Papadias and Foschini [10]:

$$\mathbf{PF4} = \begin{bmatrix} c_1 & c_2 & c_3 & c_4 \\ c_2^* & -c_1^* & c_4^* & -c_3^* \\ c_3 & -c_4 & -c_1 & c_2 \\ c_4^* & c_3^* & -c_2^* & -c_1^* \end{bmatrix}. \tag{2.22}$$

No specified construction method was given for this code.

For these three QO-STBCs, the four transmitted data symbols contained in a codeword can be separated into two groups, and its ML decoding can be performed by jointly detecting two complex symbols in each group separately, hence we call these codes *double–symbol–decodable*. We shall now use $\mathbf{J4}$ in (2.19) as an example to illustrate the groupwise ML decoding of QO-STBC.

The ML decoding metrics for $\mathbf{J4}$ have been derived in [8] and is given in (2.23). It can be seen that the decoding decisions for the symbols c_1 and c_4 are jointly obtained by minimizing the metric f_{14}, similarly the decoding decisions for symbols c_2 and c_3 are jointly obtained by minimizing the metric f_{23}. Since c_1 and c_4 (or c_2 and c_3) are each a complex symbol, their ML decoding requires the joint detection of two complex symbols (or four real symbols). Clearly, the joint detection of c_1 and c_4 can be performed independently from, and hence in parallel with, the joint detection of c_2 and c_3. This is clearly less complex than having to jointly detect all four symbols, which is the case when a non-orthogonal STBC is used.

$$
f_{14}(c_1,c_4) = \sum_{k=1}^{N_r}
\left[
\begin{array}{l}
\left(\displaystyle\sum_{i=1}^{4}|h_{i,k}|^2\right)\left(|c_1|^2+|c_4|^2\right)+ \\[2ex]
2\,\mathrm{Re}\left\{
\begin{array}{l}
c_1\left(\begin{array}{l}-h_{1,k}r_1^{(k)*}-h_{2,k}^*r_2^{(k)}\\-h_{3,k}^*r_3^{(k)}-h_{4,k}r_4^{(k)*}\end{array}\right)+ \\[3ex]
c_4\left(\begin{array}{l}-h_{4,k}r_1^{(k)*}+h_{3,k}^*r_2^{(k)}\\+h_{2,k}^*r_3^{(k)}-h_{1,k}r_4^{(k)*}\end{array}\right)+ \\[3ex]
c_1c_4^*\times 2\,\mathrm{Re}\left(\begin{array}{l}h_{1,k}h_{4,k}^*\\-h_{2,k}h_{3,k}^*\end{array}\right)
\end{array}\right\}
\end{array}
\right],
$$

$$
f_{23}(c_2,c_3) = \sum_{k=1}^{N_r}
\left[
\begin{array}{l}
\left(\displaystyle\sum_{i=1}^{4}|h_{i,k}|^2\right)\left(|c_2|^2+|c_3|^2\right)+ \\[2ex]
2\,\mathrm{Re}\left\{
\begin{array}{l}
c_2\left(\begin{array}{l}-h_{2,k}r_1^{(k)*}+h_{1,k}^*r_2^{(k)}\\-h_{4,k}^*r_3^{(k)}+h_{3,k}r_4^{(k)*}\end{array}\right)+ \\[3ex]
c_3\left(\begin{array}{l}-h_{3,k}r_1^{(k)*}-h_{4,k}^*r_2^{(k)}\\+h_{1,k}^*r_3^{(k)}+h_{2,k}r_4^{(k)*}\end{array}\right)+ \\[3ex]
c_2c_3^*\times 2\,\mathrm{Re}\left(\begin{array}{l}-h_{1,k}h_{4,k}^*\\+h_{2,k}h_{3,k}^*\end{array}\right)
\end{array}\right\}
\end{array}
\right]. \tag{2.23}
$$

Fig. 2.1 shows the open-loop capacity of a four transmit and one receive antenna system, obtained based on (1.9), as well as the achievable capacities of rate-1 QO-STBC and rate-3/4 O-STBC, obtained based on (1.18). It shows that the rate-1 QO-STBC capacity approaches the open-loop capacity of a four transmit one receive antenna system very closely, whereas the rate-3/4 O-STBC capacity does not, due to its lower code rate. Hence QO-STBC is attractive for its ability to both approach the open-loop capacity and support a lower decoding complexity than the non-orthogonal STBC.

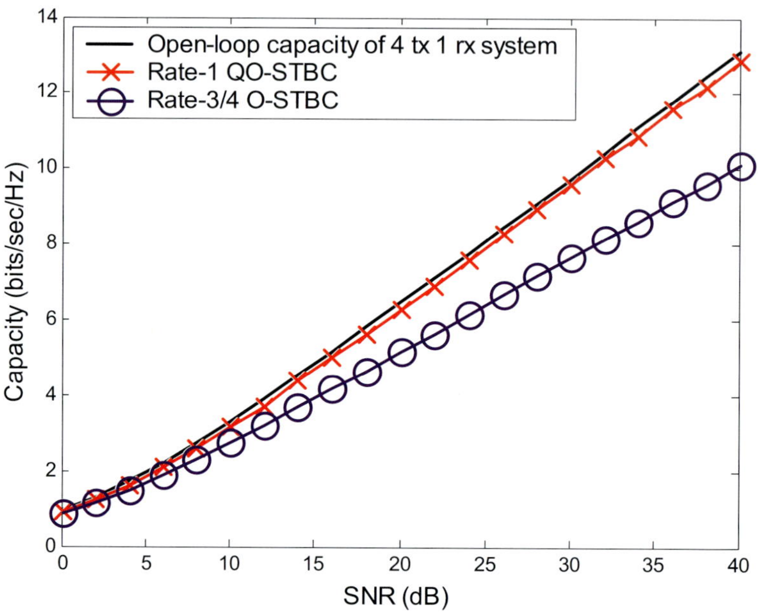

Fig. 2.1 Capacity of MISO channel with four transmit and one receive antennas

Rate-3/4 QO-STBC for eight transmit antennas

For the case of eight transmit antennas, there are two classes of QO-STBC in general. The first one is a rate-3/4 QO-STBC which transmits six complex data symbols over eight time periods. The six transmitted data symbols of this code can be separated into three groups, and its ML decoding can be performed by jointly detecting two complex symbols per group (*double–symbol–decodable*). An example of this code is the QO-STBC proposed by Jafarkhani in [8], which is denoted herein as **J8**:

$$\mathbf{J8} = \begin{bmatrix} c_1 & c_2 & c_3 & 0 & c_4 & c_5 & c_6 & 0 \\ -c_2^* & c_1^* & 0 & -c_3 & c_5^* & -c_4^* & 0 & c_6 \\ c_3^* & 0 & -c_1^* & -c_2 & -c_6^* & 0 & c_4^* & c_5 \\ 0 & -c_3^* & c_2^* & -c_1 & 0 & c_6^* & -c_5^* & c_4 \\ -c_4 & -c_5 & -c_6 & 0 & c_1 & c_2 & c_3 & 0 \\ -c_5^* & c_4^* & 0 & c_6 & -c_2^* & c_1^* & 0 & c_3 \\ c_6^* & 0 & -c_4^* & c_5 & c_3^* & 0 & -c_1^* & c_2 \\ 0 & c_6^* & -c_5^* & -c_4 & 0 & c_3^* & -c_2^* & -c_1 \end{bmatrix}. \qquad (2.24)$$

<u>Rate-1 QO-STBC for eight transmit antennas</u>

Since there is a capacity loss when the code rate is less than one, another class of QO-STBC for eight transmit antennas that achieves rate one has been proposed. This QO-STBC transmits eight complex data symbols over eight time periods. The eight transmitted data symbols can be separated into two groups, and its ML decoding can be performed by jointly detecting four complex symbols per group (*quad–symbol–decodable*). An example of this code is the QO-STBC proposed by Tirkkonen, Boariu and Hottinen in [9], which is denoted herein as **TBH8**:

$$\mathbf{TBH8} = \begin{bmatrix} \mathbf{A} & \mathbf{B} & \mathbf{C} & \mathbf{D} \\ \mathbf{B} & \mathbf{A} & \mathbf{D} & \mathbf{C} \\ \mathbf{C} & \mathbf{D} & \mathbf{A} & \mathbf{B} \\ \mathbf{D} & \mathbf{C} & \mathbf{B} & \mathbf{A} \end{bmatrix}$$

$$= \begin{bmatrix} c_1 & c_2 & c_3 & c_4 & c_5 & c_6 & c_7 & c_8 \\ -c_2^* & c_1^* & -c_4^* & c_3^* & -c_6^* & c_5^* & -c_8^* & c_7^* \\ c_3 & c_4 & c_1 & c_2 & c_7 & c_8 & c_5 & c_6 \\ -c_4^* & c_3^* & -c_2^* & c_1^* & -c_8^* & c_7^* & -c_6^* & c_5^* \\ c_5 & c_6 & c_7 & c_8 & c_1 & c_2 & c_3 & c_4 \\ -c_6^* & c_5^* & -c_8^* & c_7^* & -c_2^* & c_1^* & -c_4^* & c_3^* \\ c_7 & c_8 & c_5 & c_6 & c_3 & c_4 & c_1 & c_2 \\ -c_8^* & c_7^* & -c_6^* & c_5^* & -c_4^* & c_3^* & -c_2^* & c_1^* \end{bmatrix}. \tag{2.25}$$

Another rate-1 QO-STBC example is from Yuen, Guan, and Tjhung [31], denoted as **YGT8**:

$$\mathbf{YGT8} = \begin{bmatrix} \mathbf{A} & \mathbf{B} & \mathbf{C} & \mathbf{D} \\ -\mathbf{B}^* & \mathbf{A}^* & -\mathbf{D}^* & \mathbf{C}^* \\ -\mathbf{C}^* & -\mathbf{D}^* & \mathbf{A}^* & \mathbf{B}^* \\ \mathbf{D} & -\mathbf{C} & -\mathbf{B} & \mathbf{A} \end{bmatrix}$$

$$= \begin{bmatrix} c_1 & c_2 & c_3 & c_4 & c_5 & c_6 & c_7 & c_8 \\ -c_2^* & c_1^* & -c_4^* & c_3^* & -c_6^* & c_5^* & -c_8^* & c_7^* \\ -c_3^* & -c_4^* & c_1^* & c_2^* & -c_7^* & -c_8^* & c_5^* & c_6^* \\ c_4 & -c_3 & -c_2 & c_1 & c_8 & -c_7 & -c_6 & c_5 \\ -c_5^* & -c_6^* & -c_7^* & -c_8^* & c_1^* & c_2^* & c_3^* & c_4^* \\ c_6 & -c_5 & c_8 & -c_7 & -c_2 & c_1 & -c_4 & c_3 \\ c_7 & c_8 & -c_5 & -c_6 & -c_3 & -c_4 & c_1 & c_2 \\ -c_8^* & c_7^* & c_6^* & -c_5^* & c_4^* & -c_3^* & -c_2^* & c_1^* \end{bmatrix}. \tag{2.26}$$

2.2.2 Performance optimization of QO-STBC

Although QO-STBC can achieve higher code rates than O-STBC, it generally does not provide full transmit diversity directly. As a result, they suffer poorer uncoded decoding error probability performance than O-STBC at high SNR. Hence it is of interest to find ways to improve the decoding performance of QO-STBC.

2.2.2.1 Full-diversity QO-STBC with constellation rotation

Constellation rotation (CR) is proposed in [32,33,34] to achieve full diversity in QO-STBC with ML decoding. We shall illustrate this idea by using **J4** as an example.

To examine the decoding error probability of **J4**, we look at its codeword distance matrix $\mathbf{A_{CE}}$ as defined in (1.21). The expression of $\mathbf{A_{CE}(J4)}$ is:

$$\mathbf{A_{CE}}(\mathbf{J4}) = \begin{bmatrix} \alpha & 0 & 0 & \beta \\ 0 & \alpha & -\beta & 0 \\ 0 & -\beta & \alpha & 0 \\ \beta & 0 & 0 & \alpha \end{bmatrix}, \tag{2.27}$$

where

$$\alpha = \sum_{i=1}^{4} |\Delta_i|^2,$$
$$\beta = 2 \times \mathrm{Re}(\Delta_1 \times \Delta_4^* - \Delta_2 \times \Delta_3^*), \tag{2.28}$$

and $\Delta_i = c_i - e_i$ represents the possible code symbol errors. The determinant of the codeword distance matrix of **J4** is

$$\mathrm{Det}\left(\mathbf{A_{CE}}(\mathbf{J4})\right) = (\alpha + \beta)^2 \times (\alpha - \beta)^2$$
$$= \begin{bmatrix} \left(|\Delta_1 + \Delta_4|^2 + |\Delta_2 - \Delta_3|^2\right) \times \\ \left(|\Delta_1 - \Delta_4|^2 + |\Delta_2 + \Delta_3|^2\right) \end{bmatrix}^2. \tag{2.29}$$

Due to the summation of two $|\,.\,|^2$ terms in (2.29), the minimum value of (2.29) occurs when half of the symbols have error, i.e. either $\Delta_1 = \Delta_4 = 0$ (while Δ_2 and Δ_3 are non-zero) or $\Delta_2 = \Delta_3 = 0$ (while Δ_1 and Δ_4 are non-zero). Hence when considering the minimum determinant value, (2.29) can be simplified as:

$$\text{Min}\left[\text{Det}\left(\mathbf{A}_{\mathbf{CE}}(\mathbf{J4})\right)\right]=\left[\left(\left|\Delta_1+\Delta_4\right|^2\right)\times\left(\left|\Delta_1-\Delta_4\right|^2\right)\right]^2, \qquad (2.30)$$

assuming that $\Delta_2 = \Delta_3 = 0$.

It is easy to see from (2.30) that, when $\Delta_1 = \pm\Delta_4$, the determinant value of the codeword distance matrix of **J4** would become zero, which implies that the codeword distance matrix of **J4** would not have full rank and **J4** cannot achieve full diversity. Such a situation can easily happen due to the symmetry of conventional constellation sets such as phase shift keying (PSK) and quadrature amplitude modulation (QAM). To achieve non-zero determinant value and hence full diversity, we need to have $\Delta_1 \neq \Delta_4$. This can be achieved by using the CR technique.

For example, consider the binary phase shift keying (BPSK) constellation set {-1, 1} for all symbols c_1 to c_4 in **J4**, the possible values of Δ_i are {0, 2, -2} for all values of i, hence $|\Delta_1 \pm \Delta_4|$ can be easily be zero.

But if c_1 uses a non-rotated constellation set as shown in Fig. 2.2(a) (hence Δ_1 takes values {0, 2, -2}), while c_4 uses a rotated constellation set as shown in Fig. 2.2(b) (hence Δ_4 takes values {0, 2exp(jϑ), -2exp(jϑ)}), then a non-zero value for $|\Delta_1 \pm \Delta_4|$ can always be obtained (Δ_1 and Δ_4 cannot be both zero under the assumption of $\Delta_2 = \Delta_3 = 0$, otherwise there would not be any codeword difference). A similar BPSK CR can be applied to c_2 and c_3. Therefore, CR provides a simple and effective means to achieve full diversity in QO-STBC.

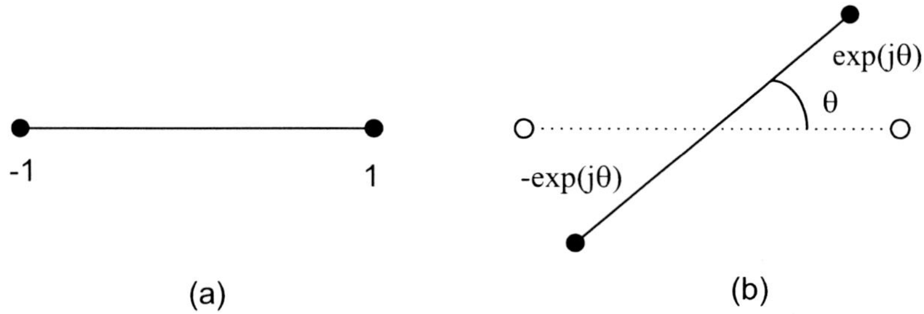

Fig. 2.2 Constellation rotation of BPSK constellation set

Besides facilitating full diversity, CR gives an extra degree of freedom to maximize the minimum determinant value in (2.30) to achieve maximum coding gain for the QO-STBC (according to the determinant criterion presented in Section 1.5 of this monograph). The **J4** code after appropriate CR, denoted as **J4_CR**, is able to achieve full diversity and optimum diversity product in (1.23) [16,31]. **J4_CR** is shown in (2.31):

$$\mathbf{J4\_CR} = \begin{bmatrix} c_1 & c_2 & c_3 e^{j\pi/4} & c_4 e^{j\pi/4} \\ -c_2^* & c_1^* & -(c_4 e^{j\pi/4})^* & (c_3 e^{j\pi/4})^* \\ -(c_3 e^{j\pi/4})^* & -(c_4 e^{j\pi/4})^* & c_1^* & c_2^* \\ c_4 e^{j\pi/4} & -c_3 e^{j\pi/4} & -c_2 & c_1 \end{bmatrix}, \quad (2.31)$$

where the factor $e^{j\pi/4}$ denotes the optimum CR angle for the QAM symbols c_3 and c_4 [16,31].

The uncoded BER performance of rate-1 QO-STBC and rate-1/2 O-STBC is shown in Fig. 2.3. In order for both codes to have the same spectral efficiency of 2 bps/Hz for fair comparison, the rate-1/2 O-STBC employs 16QAM, while the rate-1 QO-STBC employs QPSK. It can be seen from Fig. 2.3 that both **J4** and **J4_CR** perform better than O-STBC at low SNR region, as they have a higher code rate and are hence able to use QPSK that has a larger Euclidean distance than 16QAM. However, as the SNR increases, the performance of **J4** loses out, as it is not able to achieve full transmit diversity and hence its BER slope is smaller than the other two codes. On the other hand, **J4_CR**, which employs constellation rotation, performs consistently better than O-STBC. In fact, as **J4_CR** has the same BER slope as the O-STBC, we can affirm that **J4_CR** achieves full transmit diversity.

More details on the CR of QO-STBC can be found in [16,31,35,36,37].

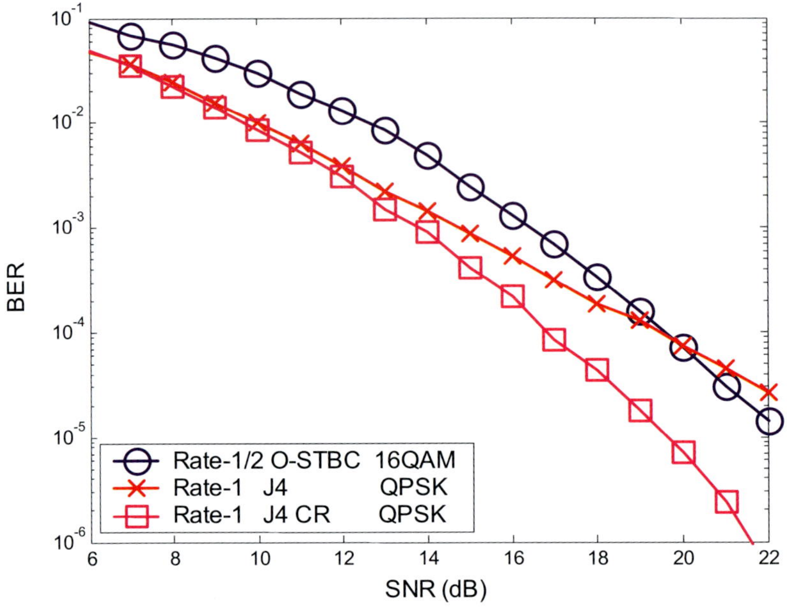

Fig. 2.3 Uncoded BER performance of QO-STBC with spectral efficiency of 2 bps/Hz

2.2.2.2 *Full-diversity QO-STBC without constellation rotation*

New QO-STBC that achieves full diversity has been proposed in [38,39,40]. When this QO-STBC is used with square-QAM, the angle of constellation rotation is either zero or $\pi/2$, hence the transmitted symbols always have the same constellation as the original square-QAM. For the case of M-PSK, the angle of constellation rotation is either zero or $2\pi/M$, hence the resultant constellation is again the same as the original M-PSK. One such example is as follows:

$$\mathbf{CPS4} = \begin{bmatrix} c_1 & c_2 & c_3 & c_4 \\ \varsigma c_2 & c_1 & \varsigma c_4 & c_3 \\ -c_3^* & -\varsigma^* c_4^* & c_1^* & \varsigma^* c_2^* \\ -c_4^* & -c_3^* & c_2^* & c_1^* \end{bmatrix}, \tag{2.32}$$

where

$$\varsigma = \begin{cases} j, & \text{for square-QAM.} \\ e^{j2\pi/M}, & \text{for M-PSK.} \end{cases} \tag{2.33}$$

The decoding performance and decoding complexity of this code has been shown to be the same as QO-STBC with CR.

2.2.3 Remark

We summarize the four classes of existing full-diversity QO-STBC in Table 2.5.

Table 2.5 Four classes of QO-STBC

QO-STBC	Transmit Antennas	Code Rate	Number of complex symbols for joint detection
J4, TBH4, PF4	4	1	2
CPS4	4	1	2
J8	8	3/4	2
TBH8, YGT8	8	1	4

Since an attractive attribute of QO-STBC is its group-wise ML decoding capability, it is natural for one to ask if the joint detection dimensions of the existing QO-STBC can be further reduced. In the next chapter, we shall present a new constellation transformation technique for the QO-STBC in the shaded rows of Table 2.5 such that the number of joint detection symbols required by them can be halved but the code can still attain full transmit diversity.

Chapter 3

Insights of QO-STBC

In this chapter, we will derive the algebraic structure of QO-STBC, and present a decoding framework for QO-STBC based on noise whitening, to facilitate the understanding of the quasi-orthogonal group concept in QO-STBC. Based on the derived QO-STBC structure, we will show that the conventional constellation rotation (CR) in fact increases the decoding complexity of the QO-STBC which it optimizes. This leads us to introduce a new constellation transformation technique, called group-constrained linear transformation (GCLT), to optimize the QO-STBC without increasing its decoding complexity.

3.1 Algebraic Structure of QO-STBC

In the idea of quasi-orthogonal STBC design in [8], the transmission matrix columns are divided into different groups. Columns in different groups are orthogonal to each other, while columns within each group are not. We will *not* use this definition of QO-STBC in this monograph. Instead, we will use the definition that the equivalent channel matrix $\mathbf{H}$ of QO-STBC, given earlier in (1.13), has the property that $\mathbf{H}^{T}\mathbf{H}$ is block diagonal. Using this new definition, we are able to separate K received symbols in a STBC codeword into G independent groups, such that symbols in any group are independent of all symbols in the other groups. ML decoding of this QO-STBC can then be performed by decoding the received symbols group by group. Specifically, the ML decoding of each group can be achieved by jointly detecting only K/G complex symbols, instead of jointly detecting all the K symbols (which is clearly a very complex operation).

In addition to achieving the above-mentioned reduction in ML decoding complexity, it is also important to ensure that the QO-STBC achieve full diversity. Hence we will divide the design of QO-STBC in this chapter into two parts, one governing the grouping of code symbols (hence governing the decoding complexity), the other governing the optimization of decoding performance.

3.1.1 Decoding complexity of a QO-STBC

Definition 3.1: A quasi-orthogonal STBC $\mathbf{C}$ is one whose equivalent channel matrix $\mathbf{H}$, as defined in (1.13), has the property that $\mathbf{H}^{\mathrm{T}}\mathbf{H}$ is block-diagonal and consists of G smaller sub-matrices each with size $(2K/G) \times (2K/G)$, or that $\mathbf{H}^{\mathrm{T}}\mathbf{H}$ can be rendered block diagonal by a permutation, i.e. $\mathbf{P}^{\mathrm{T}}\mathbf{H}^{\mathrm{T}}\mathbf{H}\mathbf{P}$ is block diagonal where $\mathbf{P}^{\mathrm{T}}\mathbf{P} = \mathbf{I}$ and $\mathbf{P}$ has only unit entries. The symbols in different sub-matrices are said to be in different groups and independent, while the symbols in the same sub-matrix are said to be in the same group and not separable.

It will be proven in *Theorem 3.2* that the ML decoding of $\mathbf{C}$ as defined in *Definition 3.1* can be performed by G independent decoders in parallel, and each of the decoders only needs to jointly detect K/G complex symbols. Therefore, the number of groups in a QO-STBC, G, determines the ML decoding complexity of the QO-STBC. A QO-STBC with larger G has less symbols for joint detection per group, hence lower ML decoding complexity.

Refering to *Definition 3.1*, O-STBC can be treated as a special case where $\mathbf{H}^{\mathrm{T}}\mathbf{H}$ is fully diagonal and $G = 2K$, hence the real and imaginary part of every symbols in O-STBC are fully separated.

Without loss of generality, throughout this monograph, we assume that $\mathbf{P} = \mathbf{I}$, where $\mathbf{I}$ is an identity matrix of appropriate dimension.

Theorem 3.1: For any two symbols in different groups (indexed using subscripts p and q) in a QO-STBC to be separable as defined in *Definition 3.1*, their dispersion matrices $\{\mathbf{A}_p, \mathbf{B}_p\}$ and $\{\mathbf{A}_q, \mathbf{B}_q\}$ defined as per (1.10) must possess the following algebraic structure, herein referred to as *Quasi-Orthogonality Constraints* (QOC):

$$\text{(i) } \mathbf{A}_p^{\mathrm{H}}\mathbf{A}_q = -\mathbf{A}_q^{\mathrm{H}}\mathbf{A}_p;$$

$$\text{(ii) } \mathbf{B}_p^{\mathrm{H}}\mathbf{B}_q = -\mathbf{B}_q^{\mathrm{H}}\mathbf{B}_p; \qquad 1 \le q \notin \mathcal{G}(p) \le K, \qquad (3.1)$$

$$\text{(iii) } \mathbf{A}_p^{\mathrm{H}}\mathbf{B}_q = \mathbf{B}_q^{\mathrm{H}}\mathbf{A}_p,$$

where $\mathcal{G}(p)$ represents a set of indexes of symbols in the same group as the symbol with index p, including p.

Proof of Theorem 3.1: To derive the QOC in (3.1), let us multiply a matched filter $(\mathbf{H}^{\mathrm{T}})$ to the received signal $\tilde{\mathbf{r}}$ in (1.13) to obtain $\mathbf{H}^{\mathrm{T}}\tilde{\mathbf{r}} = \sqrt{\rho/N_T}\,\mathbf{H}^{\mathrm{T}}\mathbf{H}\tilde{\mathbf{c}} + \mathbf{H}^{\mathrm{T}}\tilde{\boldsymbol{\eta}}$, and consider a segment of $\mathbf{H}^{\mathrm{T}}\mathbf{H}$ as shown below:

$$\mathbf{H}^{\mathrm{T}}\mathbf{H}$$

$$= \begin{bmatrix} \vdots & \cdots & \vdots \\ \tilde{\mathbf{h}}_1^{\mathrm{T}}\mathcal{A}_p^{\mathrm{T}} & \cdots & \tilde{\mathbf{h}}_{N_R}^{\mathrm{T}}\mathcal{A}_p^{\mathrm{T}} \\ \tilde{\mathbf{h}}_1^{\mathrm{T}}\mathcal{B}_p^{\mathrm{T}} & \cdots & \tilde{\mathbf{h}}_{N_R}^{\mathrm{T}}\mathcal{B}_p^{\mathrm{T}} \\ \tilde{\mathbf{h}}_1^{\mathrm{T}}\mathcal{A}_q^{\mathrm{T}} & \cdots & \tilde{\mathbf{h}}_{N_R}^{\mathrm{T}}\mathcal{A}_q^{\mathrm{T}} \\ \tilde{\mathbf{h}}_1^{\mathrm{T}}\mathcal{B}_q^{\mathrm{T}} & \cdots & \tilde{\mathbf{h}}_{N_R}^{\mathrm{T}}\mathcal{B}_q^{\mathrm{T}} \\ \vdots & \cdots & \vdots \end{bmatrix} \begin{bmatrix} \cdots & \mathcal{A}_p\tilde{\mathbf{h}}_1 & \mathcal{B}_p\tilde{\mathbf{h}}_1 & \mathcal{A}_q\tilde{\mathbf{h}}_1 & \mathcal{B}_q\tilde{\mathbf{h}}_1 & \cdots \\ & \vdots & \vdots & \vdots & \vdots & \vdots & \vdots \\ \cdots & \mathcal{A}_p\tilde{\mathbf{h}}_{N_R} & \mathcal{B}_p\tilde{\mathbf{h}}_{N_R} & \mathcal{A}_q\tilde{\mathbf{h}}_{N_R} & \mathcal{B}_q\tilde{\mathbf{h}}_{N_R} & \cdots \end{bmatrix}, \; 1 \le q \notin \mathcal{G}(p) \le K$$

$$= \begin{bmatrix} \ddots & \cdots & \cdots & \cdots & \cdots & \\ \cdots & \sum_{i=1}^{N_R}\tilde{\mathbf{h}}_i^{\mathrm{T}}(\mathcal{A}_p^{\mathrm{T}}\mathcal{A}_p)\tilde{\mathbf{h}}_i & \sum_{i=1}^{N_R}\tilde{\mathbf{h}}_i^{\mathrm{T}}(\mathcal{A}_p^{\mathrm{T}}\mathcal{B}_p)\tilde{\mathbf{h}}_i & \boxed{\sum_{i=1}^{N_R}\tilde{\mathbf{h}}_i^{\mathrm{T}}(\mathcal{A}_p^{\mathrm{T}}\mathcal{A}_q)\tilde{\mathbf{h}}_i \quad \sum_{i=1}^{N_R}\tilde{\mathbf{h}}_i^{\mathrm{T}}(\mathcal{A}_p^{\mathrm{T}}\mathcal{B}_q)\tilde{\mathbf{h}}_i} & \cdots \\ \cdots & \sum_{i=1}^{N_R}\tilde{\mathbf{h}}_i^{\mathrm{T}}(\mathcal{B}_p^{\mathrm{T}}\mathcal{A}_p)\tilde{\mathbf{h}}_i & \sum_{i=1}^{N_R}\tilde{\mathbf{h}}_i^{\mathrm{T}}(\mathcal{B}_p^{\mathrm{T}}\mathcal{B}_p)\tilde{\mathbf{h}}_i & \boxed{\sum_{i=1}^{N_R}\tilde{\mathbf{h}}_i^{\mathrm{T}}(\mathcal{B}_p^{\mathrm{T}}\mathcal{A}_q)\tilde{\mathbf{h}}_i \quad \sum_{i=1}^{N_R}\tilde{\mathbf{h}}_i^{\mathrm{T}}(\mathcal{B}_p^{\mathrm{T}}\mathcal{B}_q)\tilde{\mathbf{h}}_i} & \cdots \\ \cdots & \boxed{\sum_{i=1}^{N_R}\tilde{\mathbf{h}}_i^{\mathrm{T}}(\mathcal{A}_q^{\mathrm{T}}\mathcal{A}_p)\tilde{\mathbf{h}}_i \quad \sum_{i=1}^{N_R}\tilde{\mathbf{h}}_i^{\mathrm{T}}(\mathcal{A}_q^{\mathrm{T}}\mathcal{B}_p)\tilde{\mathbf{h}}_i} & & \sum_{i=1}^{N_R}\tilde{\mathbf{h}}_i^{\mathrm{T}}(\mathcal{A}_q^{\mathrm{T}}\mathcal{A}_q)\tilde{\mathbf{h}}_i & \sum_{i=1}^{N_R}\tilde{\mathbf{h}}_i^{\mathrm{T}}(\mathcal{A}_q^{\mathrm{T}}\mathcal{B}_q)\tilde{\mathbf{h}}_i & \cdots \\ \cdots & \boxed{\sum_{i=1}^{N_R}\tilde{\mathbf{h}}_i^{\mathrm{T}}(\mathcal{B}_q^{\mathrm{T}}\mathcal{A}_p)\tilde{\mathbf{h}}_i \quad \sum_{i=1}^{N_R}\tilde{\mathbf{h}}_i^{\mathrm{T}}(\mathcal{B}_q^{\mathrm{T}}\mathcal{B}_p)\tilde{\mathbf{h}}_i} & & \sum_{i=1}^{N_R}\tilde{\mathbf{h}}_i^{\mathrm{T}}(\mathcal{B}_q^{\mathrm{T}}\mathcal{A}_q)\tilde{\mathbf{h}}_i & \sum_{i=1}^{N_R}\tilde{\mathbf{h}}_i^{\mathrm{T}}(\mathcal{B}_q^{\mathrm{T}}\mathcal{B}_q)\tilde{\mathbf{h}}_i & \cdots \\ & \cdots & \cdots & \cdots & \cdots & \ddots \end{bmatrix}$$

$$(3.2)$$

In order to achieve separation between two symbols in different groups, e.g. between symbols c_p and c_q, the summation terms included in the enclosed boxes in (3.2) are required to be zero. Note that these summation terms have the form of $\mathbf{v}^{\mathrm{T}}\mathbf{M}\mathbf{v}$. For any vector $\mathbf{v}$ of size $v \times 1$, $\mathbf{v}^{\mathrm{T}}\mathbf{M}\mathbf{v} = 0$ if and only if the matrix $\mathbf{M}$ of size $v \times v$ is skew-symmetric, i.e. $\mathbf{M}^{\mathrm{T}} = -\mathbf{M}$. Hence $\mathcal{A}_p^{\mathrm{T}}\mathcal{A}_q$, $\mathcal{A}_p^{\mathrm{T}}\mathcal{B}_q$, $\mathcal{B}_p^{\mathrm{T}}\mathcal{A}_q$, and $\mathcal{B}_p^{\mathrm{T}}\mathcal{B}_q$ have to be skew-symmetric for p and q that are of different group.

Next, from (3.1) (i), we get

$$\mathbf{A}_p^{\mathrm{H}}\mathbf{A}_q = -\mathbf{A}_q^{\mathrm{H}}\mathbf{A}_p$$

$$\Rightarrow (\mathbf{A}_p^{\mathrm{R}} + j\mathbf{A}_p^{\mathrm{I}})^{\mathrm{H}}(\mathbf{A}_q^{\mathrm{R}} + j\mathbf{A}_q^{\mathrm{I}}) = -(\mathbf{A}_q^{\mathrm{R}} + j\mathbf{A}_q^{\mathrm{I}})^{\mathrm{H}}(\mathbf{A}_p^{\mathrm{R}} + j\mathbf{A}_p^{\mathrm{I}}) \qquad (3.3)$$

$$\Rightarrow \begin{cases} \text{real part}: \left(\mathbf{A}_p^{\mathrm{R}}\right)^{\mathrm{T}}\mathbf{A}_q^{\mathrm{R}} + \left(\mathbf{A}_p^{\mathrm{I}}\right)^{\mathrm{T}}\mathbf{A}_q^{\mathrm{I}} = -\left(\mathbf{A}_q^{\mathrm{R}}\right)^{\mathrm{T}}\mathbf{A}_p^{\mathrm{R}} - \left(\mathbf{A}_q^{\mathrm{I}}\right)^{\mathrm{T}}\mathbf{A}_p^{\mathrm{I}}; \\ \text{imag part}: \left(\mathbf{A}_p^{\mathrm{R}}\right)^{\mathrm{T}}\mathbf{A}_q^{\mathrm{I}} - \left(\mathbf{A}_p^{\mathrm{I}}\right)^{\mathrm{T}}\mathbf{A}_q^{\mathrm{R}} = \left(\mathbf{A}_q^{\mathrm{I}}\right)^{\mathrm{T}}\mathbf{A}_p^{\mathrm{R}} - \left(\mathbf{A}_q^{\mathrm{R}}\right)^{\mathrm{T}}\mathbf{A}_p^{\mathrm{I}}. \end{cases}$$

Defining

$$\mathbf{M} \triangleq \mathrm{Re}\left(\mathbf{A}_p^{\mathrm{H}}\mathbf{A}_q\right) = \left(\mathbf{A}_p^{\mathrm{R}}\right)^{\mathrm{T}}\mathbf{A}_q^{\mathrm{R}} + \left(\mathbf{A}_p^{\mathrm{I}}\right)^{\mathrm{T}}\mathbf{A}_q^{\mathrm{I}},$$

$$\mathbf{N} \triangleq \mathrm{Im}\left(\mathbf{A}_p^{\mathrm{H}}\mathbf{A}_q\right) = \left(\mathbf{A}_p^{\mathrm{R}}\right)^{\mathrm{T}}\mathbf{A}_q^{\mathrm{I}} - \left(\mathbf{A}_p^{\mathrm{I}}\right)^{\mathrm{T}}\mathbf{A}_q^{\mathrm{R}}, \qquad (3.4)$$

then (3.3) shows that $\mathbf{M}$ is skew-symmetric (i.e. $\mathbf{M}^{\mathrm{T}} = -\mathbf{M}$) and $\mathbf{N}$ is symmetric (i.e. $\mathbf{N}^{\mathrm{T}} = \mathbf{N}$). From the definition of $\mathcal{A}_q$ in (1.13), we know that

$$\mathcal{A}_p^{\mathrm{T}}\mathcal{A}_q = \begin{bmatrix} \mathbf{M} & -\mathbf{N} \\ \mathbf{N} & \mathbf{M} \end{bmatrix}, \qquad (3.5)$$

hence it is skew-symmetric. Similar conclusions can be drawn on $\mathcal{A}_p^{\mathrm{T}}\mathcal{B}_q$, $\mathcal{B}_p^{\mathrm{T}}\mathcal{A}_q$, and $\mathcal{B}_p^{\mathrm{T}}\mathcal{B}_q$.

In summary, by restricting the dispersion matrices $\{\mathbf{A}, \mathbf{B}\}$ of symbols in different groups in a QO-STBC to satisfy the QOC in (3.1), the $\mathcal{A}_p^{\mathrm{T}}\mathcal{A}_q$, $\mathcal{A}_p^{\mathrm{T}}\mathcal{B}_q$, $\mathcal{B}_p^{\mathrm{T}}\mathcal{A}_q$, and $\mathcal{B}_p^{\mathrm{T}}\mathcal{B}_q$ matrices of the QO-STBC are skew symmetric. This makes the $\mathbf{H}^{\mathrm{T}}\mathbf{H}$ matrix of the QO-STBC block diagonal, and symbols in different groups (sub-matrices of $\mathbf{H}^{\mathrm{T}}\mathbf{H}$) separable from each other. Hence *Theorem 3.1* is proved. ∎

3.1.2 Maximal symbol-wise diversity of a QO-STBC

The QOC in *Theorem 3.1* only specifies the decoupling of symbols during the decoding process of a QO-STBC; but it does not guarantee good decoding performance of the QO-STBC. To ensure that the QO-STBC can achieve full diversity, we introduce an additional constraint on the QO-STBC code structure based on the maximal symbol-wise diversity (MSD) requirement from [11,13]:

$$\mathbf{A}_i^{H}\mathbf{A}_i = \mathbf{B}_i^{H}\mathbf{B}_i = \left(\frac{P}{K}\right)\mathbf{I}_{N_T}, \qquad 1 \le i \le K, \qquad (3.6)$$

where the factor P/K is for normalizing the total transmission power. By (3.6), we guarantee full diversity protection for the most probable one-symbol error events, which has the smallest Euclidean distances for every code symbol [11,13]. To further protect against multiple-symbol error events so as to provide full diversity for all error events, constellation rotation (CR) or group-constrained linear transformation (GCLT, which will be elaborated later in Section 3.3) can next be applied.

It can be shown that all the QO-STBCs proposed in the literature, such as those in [8,9,10], comply with the algebraic structure QOC (3.1) and MSD constraint (3.6). In addition, the condition (2.2) (i) of O-STBC is exactly the MSD constraint.

3.2 Generalized Decoding Framework of QO-STBC

In the literature, researchers have considered different types of decoding schemes on different QO-STBCs. For example, in [8], the ML decision metric of the QO-STBC **J4** in (2.19) was derived. In [9], a least-mean-square estimation was used for decoding the QO-STBC **TBH4** in (2.21). In [10], the open-loop channel capacities of the QO-STBC **PF4** in (2.22) with zero-forcing (ZF), minimum mean square error (MMSE) and ML decoding were analyzed. In [41], the diversity properties of ZF and MMSE linear receivers for the QO-STBC proposed in [42] were studied. In [43,44], sphere decoding (SD) was investigated for the QO-STBC **J4**. These decoding schemes are code-specific and cannot be generalized to other QO-STBCs. To our knowledge, there has been no study on a generalised decoding approach for QO-STBC. This makes it difficult to have a unified concept of decoding complexity. In this section, we will develop a generalized decoding framework for QO-STBC to serve as the basis of group-wise ML decoding in QO-STBC.

Noise Whitening Filter for QO-STBC

The received signal $\tilde{\mathbf{r}}$ of (1.13) after matched filtering by $\mathbf{H}^T$ can be written as

$$\mathbf{H}^T\tilde{\mathbf{r}} = \sqrt{\rho/N_T}\,\mathbf{H}^T\mathbf{H}\tilde{\mathbf{c}} + \mathbf{H}^T\tilde{\boldsymbol{\eta}}$$
$$\Rightarrow \tilde{\mathbf{r}}_c = \mathbf{H}_c\tilde{\mathbf{c}} + \tilde{\boldsymbol{\eta}}_c, \tag{3.7}$$

where $\tilde{\mathbf{r}}_c = \mathbf{H}^T\tilde{\mathbf{r}}$, $\mathbf{H}_c = \sqrt{\rho/N_T}\,\mathbf{H}^T\mathbf{H}$, $\tilde{\boldsymbol{\eta}}_c = \mathbf{H}^T\tilde{\boldsymbol{\eta}}$.

Note that after the matched filtering in (3.7), $\tilde{\boldsymbol{\eta}}_c$ is a colored noise because it has covariance matrix $\mathrm{E}\left[\tilde{\boldsymbol{\eta}}_c\tilde{\boldsymbol{\eta}}_c^H\right] = N_o^2\mathbf{H}^T\mathbf{H}$ and $\mathbf{H}^T\mathbf{H}$ is not a unitary matrix. This colored-noise system is shown in Fig. 3.1. Hence traditional equalizer derived based on white noise assumption cannot be directly applied to decode the matched filter output $\tilde{\mathbf{r}}_c$, or else the optimal performance cannot be achieved. In this section, we first whiten the noise in $\tilde{\mathbf{r}}_c$, so that standard ML decoding, sphere decoding, or equalization schemes can be applied to decode the whitened received signals. This alleviates the need to design customized colored-noise solutions on a case-by-case basis.

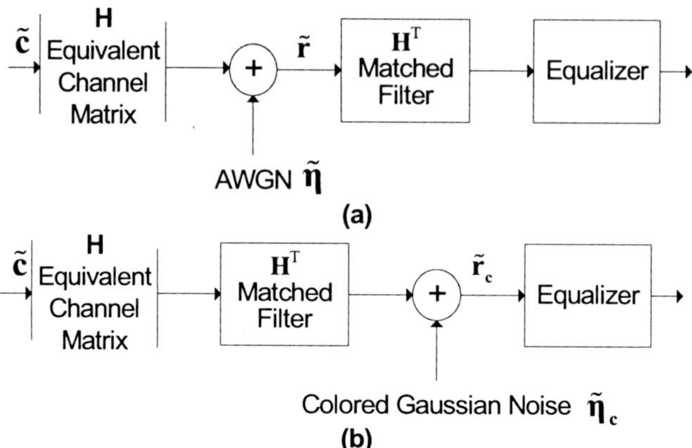

Fig. 3.1 Model of QO-STBC receiver with only matched filtering

As shown in Fig. 3.2, a whitening filter $\mathbf{H_W}$ can be employed to "whiten" the noise before the matched-filter output signal is processed by the equalizer. The resultant whitened noise $\tilde{\boldsymbol{\eta}}_w = \mathbf{H}_w \tilde{\boldsymbol{\eta}}_c = \mathbf{H}_w \mathbf{H}^T \tilde{\boldsymbol{\eta}}$ has covariance matrix $E\left[\tilde{\boldsymbol{\eta}}_w \tilde{\boldsymbol{\eta}}_w^H\right] = N_o^2 \mathbf{H}_w \mathbf{H}^T \mathbf{H} \mathbf{H}_w^T$. This implies that we should choose $\mathbf{H}_w$ to obtain $\mathbf{H}_w \mathbf{H}^T \mathbf{H} \mathbf{H}_w^T = \mathbf{I}$. This gives a solution of $\mathbf{H}_w = \left(\mathbf{H}^T \mathbf{H}\right)^{-1/2}$. Since $\mathbf{H}^T \mathbf{H}$ is symmetric, it can be decomposed according to Symmetric Schur Decomposition [46] as follows:

$$\mathbf{H}^T\mathbf{H} = \mathbf{Q}\boldsymbol{\Lambda}\mathbf{Q}^T = \sum_{i=1}^{m} \lambda_i \mathbf{q}_i \mathbf{q}_i^T , \qquad (3.8)$$

where $\mathbf{Q}\mathbf{Q}^T=\mathbf{I}$, $\boldsymbol{\Lambda}=\mathrm{diag}(\lambda_1, \lambda_2, \ldots \lambda_m)$, and m, λ_i and $\mathbf{q}_i$ $(1 \leq i \leq m)$ are the rank, eigenvalues and eigenvectors of $\mathbf{H}^T\mathbf{H}$ respectively. So the whitening filter can be shown to be:

$$\mathbf{H_W} = \left(\mathbf{H}^T\mathbf{H}\right)^{-1/2}$$
$$= \sum_{i=1}^{m} \lambda_i^{-1/2} \mathbf{q}_i \mathbf{q}_i^T . \qquad (3.9)$$

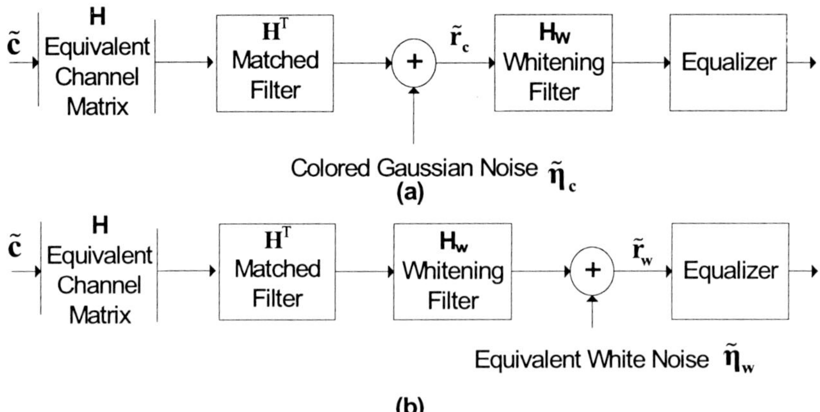

Fig. 3.2 Model of QO-STBC receiver with noise whitening

As shown in Fig. 3.3, we can now combine the equivalent channel matrix $\mathbf{H}$, the matched filter $\mathbf{H}^T$ and the whitening filter $\mathbf{H_W}$ into an equivalent filter $\mathbf{H_e}$:

$$
\begin{aligned}
\mathbf{H_e} &= \mathbf{H_W}\mathbf{H}^T\mathbf{H} \\
&= \left(\mathbf{H}^T\mathbf{H}\right)^{-1/2}\left(\mathbf{H}^T\mathbf{H}\right) \\
&= \left(\mathbf{H}^T\mathbf{H}\right)^{1/2} \\
&= \sum_{i=1}^{m}\lambda_i^{1/2}\mathbf{q}_i\mathbf{q}_i^T .
\end{aligned}
\tag{3.10}
$$

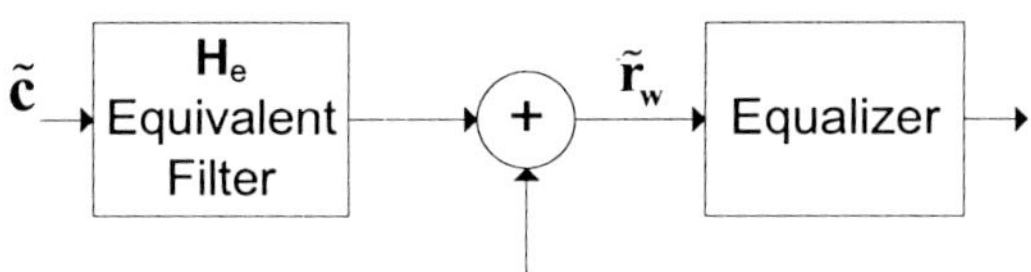

Fig. 3.3 Model of QO-STBC receiver with noise whitening

From (3.10), it can be noted that $\mathbf{H_e}$ has a similar structure as $\mathbf{H^T H}$ in (3.8), and the eigenvalues of $\mathbf{H_e}$ is the square root of the eigenvalues of $\mathbf{H^T H}$. Finally, referring to Fig. 3.3, the whitened signal $\tilde{\mathbf{r}}_{\mathbf{w}}$ can be modeled as:

$$\tilde{\mathbf{r}}_{\mathbf{w}} = \mathbf{H_w H^T \tilde{r}}$$
$$= \sqrt{\rho/N_T}\,\mathbf{H_w H^T H \tilde{c}} + \mathbf{H_w H^T \tilde{\eta}} \qquad (3.11)$$
$$= \sqrt{\rho/N_T}\,\mathbf{H_e \tilde{c}} + \tilde{\eta}_{\mathbf{w}}.$$

Theorem 3.2: The equivalent filter $\mathbf{H_e}$ of a QO-STBC has the same block diagonal structure as $\mathbf{H^T H}$. Hence the application of noise whitening filter will not alter the decoding complexity of a QO-STBC, i.e. the decoding (including ML decoding) of a QO-STBC defined in *Definition 3.1* can be performed by G independent sub-decoders, and each of the decoders only needs to jointly detect K/G complex symbols.

Proof of Theorem 3.2: One can assume the block diagonal matrix $\mathbf{H^T H}$ of a QO-STBC as follows:

$$\mathbf{H^T H} = \begin{bmatrix} \mathbf{M}_1 & \mathbf{0} & \mathbf{0} & \mathbf{0} \\ \mathbf{0} & \mathbf{M}_2 & \mathbf{0} & \mathbf{0} \\ \mathbf{0} & \mathbf{0} & \ddots & \mathbf{0} \\ \mathbf{0} & \mathbf{0} & \mathbf{0} & \mathbf{M}_G \end{bmatrix}. \qquad (3.12)$$

Since $\mathbf{H^T H}$ is symmetry, i.e. $\mathbf{M}_i$ (for $1 \leq i \leq G$) are all symmetric matrices, so each of them can be expressed as:

$$\mathbf{M}_i^T \mathbf{M}_i = \mathbf{Q}_{M_i} \mathbf{\Lambda}_{M_i} \mathbf{Q}_{M_i}^T. \qquad (3.13)$$

From (3.8), (3.12), and (3.13), it can be shown that:

$$\mathbf{Q} = \begin{bmatrix} \mathbf{Q}_{M_1} & \mathbf{0} & \mathbf{0} & \mathbf{0} \\ \mathbf{0} & \mathbf{Q}_{M_2} & \mathbf{0} & \mathbf{0} \\ \mathbf{0} & \mathbf{0} & \ddots & \mathbf{0} \\ \mathbf{0} & \mathbf{0} & \mathbf{0} & \mathbf{Q}_{M_G} \end{bmatrix},$$

$$\mathbf{\Lambda} = \begin{bmatrix} \mathbf{\Lambda}_{M_1} & \mathbf{0} & \mathbf{0} & \mathbf{0} \\ \mathbf{0} & \mathbf{\Lambda}_{M_2} & \mathbf{0} & \mathbf{0} \\ \mathbf{0} & \mathbf{0} & \ddots & \mathbf{0} \\ \mathbf{0} & \mathbf{0} & \mathbf{0} & \mathbf{\Lambda}_{M_G} \end{bmatrix}.$$

(3.14)

From (3.10) and the above,

$$\mathbf{H}_{\mathbf{e}} = \left(\mathbf{H}^{\mathrm{T}} \mathbf{H} \right)^{1/2}$$

$$= \mathbf{Q}\sqrt{\mathbf{\Lambda}}\mathbf{Q}^{\mathrm{T}}$$

$$= \begin{bmatrix} \mathbf{Q}_{M_1} & \mathbf{0} & \mathbf{0} & \mathbf{0} \\ \mathbf{0} & \mathbf{Q}_{M_2} & \mathbf{0} & \mathbf{0} \\ \mathbf{0} & \mathbf{0} & \ddots & \mathbf{0} \\ \mathbf{0} & \mathbf{0} & \mathbf{0} & \mathbf{Q}_{M_G} \end{bmatrix} \begin{bmatrix} \sqrt{\mathbf{\Lambda}_{M_1}} & \mathbf{0} & \mathbf{0} & \mathbf{0} \\ \mathbf{0} & \sqrt{\mathbf{\Lambda}_{M_2}} & \mathbf{0} & \mathbf{0} \\ \mathbf{0} & \mathbf{0} & \ddots & \mathbf{0} \\ \mathbf{0} & \mathbf{0} & \mathbf{0} & \sqrt{\mathbf{\Lambda}_{M_G}} \end{bmatrix} \begin{bmatrix} \mathbf{Q}_{M_1} & \mathbf{0} & \mathbf{0} & \mathbf{0} \\ \mathbf{0} & \mathbf{Q}_{M_2} & \mathbf{0} & \mathbf{0} \\ \mathbf{0} & \mathbf{0} & \ddots & \mathbf{0} \\ \mathbf{0} & \mathbf{0} & \mathbf{0} & \mathbf{Q}_{M_G} \end{bmatrix}^{\mathrm{T}}$$

$$= \begin{bmatrix} \mathbf{M}_1' & \mathbf{0} & \mathbf{0} & \mathbf{0} \\ \mathbf{0} & \mathbf{M}_2' & \mathbf{0} & \mathbf{0} \\ \mathbf{0} & \mathbf{0} & \ddots & \mathbf{0} \\ \mathbf{0} & \mathbf{0} & \mathbf{0} & \mathbf{M}_G' \end{bmatrix}.$$

(3.15)

Since the equivalent channel matrix $\mathbf{H}_{\mathbf{e}}$ is also block diagonal with G sub-matrices ($\mathbf{M}_1', \mathbf{M}_2', \cdots, \mathbf{M}_G'$), the whitened received signal $\tilde{\mathbf{r}}_{\mathbf{w}}$ in (3.11) can be separated into G sub-vectors, and the decoding of each sub-block can be performed independently. This proves that the whitening filter will not alter the decoding complexity of a QO-STBC, and the decoding (including ML decoding) of a QO-STBC can be separately performed by G independent sub-decoders.   ∎

Take the QO-STBC codeword **J4** in (2.19) as an example. According to [8], **J4** has two quasi-orthogonal groups hence its $\tilde{\mathbf{r}}_{\mathbf{w}}$ in (3.16) can be separated into the two independent parts shown in (3.17).

$$\tilde{\mathbf{r}}_{\mathbf{w}} = \mathbf{H}_e \tilde{\mathbf{c}} + \tilde{\boldsymbol{\eta}}_{\mathbf{w}}$$

$$\Rightarrow \begin{bmatrix} \tilde{\mathbf{r}}_{\mathbf{w}_1} \\ \tilde{\mathbf{r}}_{\mathbf{w}_2} \end{bmatrix} = \begin{bmatrix} \mathbf{M}_1' & \mathbf{0} \\ \mathbf{0} & \mathbf{M}_2' \end{bmatrix} \tilde{\mathbf{c}} + \begin{bmatrix} \tilde{\boldsymbol{\eta}}_{\mathbf{w}_1} \\ \tilde{\boldsymbol{\eta}}_{\mathbf{w}_2} \end{bmatrix}. \tag{3.16}$$

$$\tilde{\mathbf{r}}_{\mathbf{w}_1} = \mathbf{M}_1' \begin{bmatrix} c_1^R \\ c_1^I \\ c_4^R \\ c_4^I \end{bmatrix} + \tilde{\boldsymbol{\eta}}_{\mathbf{w}_1} = \mathbf{M}_1' \mathbf{c}_1' + \tilde{\boldsymbol{\eta}}_{\mathbf{w}_1},$$

$$\tilde{\mathbf{r}}_{\mathbf{w}_2} = \mathbf{M}_2' \begin{bmatrix} c_2^R \\ c_2^I \\ c_3^R \\ c_3^I \end{bmatrix} + \tilde{\boldsymbol{\eta}}_{\mathbf{w}_2} = \mathbf{M}_2' \mathbf{c}_2' + \tilde{\boldsymbol{\eta}}_{\mathbf{w}_2}, \tag{3.17}$$

where $\tilde{\mathbf{c}} = \begin{bmatrix} c_1^R & c_1^I & c_4^R & c_4^I & c_2^R & c_2^I & c_3^R & c_3^I \end{bmatrix}^T$ are the four complex symbols transmitted in **J4**, $\mathbf{M}_1'$ and $\mathbf{M}_2'$ are the sub-matrices of the equivalent filter of **J4**, and $\tilde{\boldsymbol{\eta}}_{\mathbf{w}_1}$ and $\tilde{\boldsymbol{\eta}}_{\mathbf{w}_2}$ are whitened noise vectors.

Each part in (3.17) contains distinct halves of the symbols in $\tilde{\mathbf{c}}$, this simplifies the decoding process and makes parallel decoding possible. For simplicity, each of the decoupled equations in (3.17) can be expressed in a general form as shown below:

$$\mathbf{r}' = \mathbf{M}' \mathbf{c}' + \boldsymbol{\eta}', \tag{3.18}$$

where $\mathbf{r}' = \tilde{\mathbf{r}}_{\mathbf{w}_1}$ or $\tilde{\mathbf{r}}_{\mathbf{w}_2}$, $\mathbf{M}' = \mathbf{M}_1'$ or $\mathbf{M}_2'$, $\mathbf{c}' = \mathbf{c}_1'$ or $\mathbf{c}_2'$, and $\boldsymbol{\eta}' = \tilde{\boldsymbol{\eta}}_{\mathbf{w}_1}$ or $\tilde{\boldsymbol{\eta}}_{\mathbf{w}_2}$.

<u>Remarks</u>

By comparing the received STBC signal in (1.13) and the QO-STBC signal with the noise whitening in (3.18), we can see that $\tilde{\mathbf{c}}$ in (1.13) is of dimension $2K \times 1$, whereas the $\mathbf{c}'$ in (3.18) is of dimension $2K/G \times 1$. Hence the matrices in (3.18) have smaller dimensions than those in (1.13) because the QO-STBC symbols in different groups are decoupled and independent. Such reduction in matrix dimension clearly leads to simplification of the QO-STBC decoding operations.

We have therefore shown that QO-STBC can be group-wise decoded, regardless of any specific decoder structure.

3.3 Impact of Constellation Rotation on the Decoding Complexity of QO-STBC

As explained in Section 2.2.2, QO-STBCs can achieve full transmit diversity with the help of CR [16,31,32,33,35]. To date, full-rate full-diversity QO-STBC for four transmit antennas requires joint detection of at least two complex symbols [16,32,33]. For eight transmit antennas, full-diversity QO-STBC requires joint detection of at least two complex symbols at a code rate of 3/4 [16], or four complex symbols at full rate [31]. These have been summarized in Table 2.5.

In this section, we shall show that the number of joint detection symbols required by the existing full-diversity QO-STBCs with square or rectangular-QAM constellations can actually be halved if, instead of CR, a different constellation transformation scheme called *"Group-Constrained Linear Transformation* (GCLT)" [47,48] is used to optimize the original QO-STBCs. To explain the principles of GCLT, we will examine the algebraic structure of the existing QO-STBCs and study the impact of CR on their decoding complexity. We will then re-optimize the full-rate QO-STBC **J4** in (2.19), rate-3/4 QO-STBC **J8** in (2.24) and rate-1 QO-STBC **TBH8** in (2.25) using GCLT. The bit error rate (BER) performance and decoding complexity of these QO-STBC optimized using CR and GCLT will finally be compared.

3.3.1 Simplified QO-STBC model with real symbols only

Since a square- or rectangular-QAM symbol can be treated as two independent real PAM symbols, we may simplify the signal model of STBC in (1.10) as follows:

$$\begin{aligned}
\mathbf{C} &= \sum_{i=1}^{K}\left(c_i^{\mathrm{R}}\mathbf{A}_i + jc_i^{\mathrm{I}}\mathbf{B}_i\right) \\
&= \sum_{i=1}^{K}(d_i\mathbf{A}_i + d_{K+i}\mathbf{A}_{K+i}) \quad \text{where } \mathbf{A}_{k+p} = j\mathbf{B}_p \quad \forall 1 \le p \le K \quad (3.19) \\
&= \sum_{k=1}^{2K}(d_k\mathbf{A}_k),
\end{aligned}$$

where the complex data symbols are re-modeled as $c_i = d_i + jd_{K+i}$ for $1 \le i \le K$. Hence transmitting K regular square- or rectangular-QAM symbols is equivalent to transmitting $2K$ PAM symbols. Accordingly, the K sets of $\{\mathbf{A}, \mathbf{B}\}$ dispersion matrices (each for a different complex symbol c_i) of a STBC are mapped to $2K$ sets of $\{\mathbf{A}\}$ dispersion matrices (each for a different real symbol d_k) in (3.19). With such modeling, the QOC in (3.1) can be simplified as follows:

Proposition 3.1: For the real symbol d_p defined in the simplified STBC model in (3.19) to be separated / decoupled from another real symbol d_q in the same STBC, their dispersion matrices $\mathbf{A}_p$ and $\mathbf{A}_q$ must follow the following version of *Quasi-Orthogonality Constraint* (QOC):

$$\mathbf{A}_p^{\mathrm{H}}\mathbf{A}_q = -\mathbf{A}_q^{\mathrm{H}}\mathbf{A}_p, \qquad \text{for } 1 \le p,q \le 2K \text{ and } q \notin \mathcal{G}(p). \quad (3.20)$$

Proof of Proposition 3.1: (3.20) can be readily proved by setting $\mathbf{A}_{K+i} = j\mathbf{B}_i$ for $1 \le i \le K$ in (3.1).   ■

In this section, we will adopt the simplified STBC signal model in (3.19) and simplified QOC in (3.20) for the discussions of GCLT. The QO-STBC examples considered will be the double–symbol–decodable **J4** in (2.19), **J8** code in (2.24) and quad–symbol–decodable **TBH8** in (2.25). These QO-STBCs are representative of most QO-STBCs reported in the literature.

3.3.2 *Decoding complexity of QO-STBC with CR*

Based on the QOC in (3.20), we now examine the quasi-orthogonality relationship between the eight dispersion matrices of the code **J4** [8] before CR (listed in (3.21) as matrices $\mathbf{A}_1$, $\mathbf{A}_2$, ..., $\mathbf{A}_8$):

$$
\mathbf{A}_1 = \begin{bmatrix} 1 & 0 & 0 & 0 \\ 0 & 1 & 0 & 0 \\ 0 & 0 & 1 & 0 \\ 0 & 0 & 0 & 1 \end{bmatrix}, \quad
\mathbf{A}_2 = \begin{bmatrix} 0 & 1 & 0 & 0 \\ -1 & 0 & 0 & 0 \\ 0 & 0 & 0 & 1 \\ 0 & 0 & -1 & 0 \end{bmatrix},
$$

$$
\mathbf{A}_3 = \begin{bmatrix} 0 & 0 & 1 & 0 \\ 0 & 0 & 0 & 1 \\ -1 & 0 & 0 & 0 \\ 0 & -1 & 0 & 0 \end{bmatrix}, \quad
\mathbf{A}_4 = \begin{bmatrix} 0 & 0 & 0 & 1 \\ 0 & 0 & -1 & 0 \\ 0 & -1 & 0 & 0 \\ 1 & 0 & 0 & 0 \end{bmatrix},
$$

$$
\mathbf{A}_5 = \begin{bmatrix} j & 0 & 0 & 0 \\ 0 & -j & 0 & 0 \\ 0 & 0 & -j & 0 \\ 0 & 0 & 0 & j \end{bmatrix}, \quad
\mathbf{A}_6 = \begin{bmatrix} 0 & j & 0 & 0 \\ j & 0 & 0 & 0 \\ 0 & 0 & 0 & -j \\ 0 & 0 & -j & 0 \end{bmatrix},
$$

$$
\mathbf{A}_7 = \begin{bmatrix} 0 & 0 & j & 0 \\ 0 & 0 & 0 & -j \\ j & 0 & 0 & 0 \\ 0 & -j & 0 & 0 \end{bmatrix}, \quad
\mathbf{A}_8 = \begin{bmatrix} 0 & 0 & 0 & j \\ 0 & 0 & j & 0 \\ 0 & j & 0 & 0 \\ j & 0 & 0 & 0 \end{bmatrix}. \tag{3.21}
$$

The fulfillment of QOC of $\mathbf{A}_1$, $\mathbf{A}_2$, ..., $\mathbf{A}_8$ is shown in Table 3.1(a). For example, $\mathbf{A}_1$ satisfies QOC (3.20) with all other dispersion matrices except $\mathbf{A}_4$. Likewise, each of $\mathbf{A}_2$, ..., $\mathbf{A}_8$ satisfies QOC (3.20) with all but one of the other dispersion matrices. By re-arranging the rows and columns of Table 3.1(a), we obtain Table 3.1(b), which shows that the dispersion matrices $\mathbf{A}_1$, $\mathbf{A}_2$, ..., $\mathbf{A}_8$ of the code **J4** can be grouped into four orthogonal groups: $\{\mathbf{A}_1, \mathbf{A}_4\}$, $\{\mathbf{A}_2, \mathbf{A}_3\}$, $\{\mathbf{A}_5, \mathbf{A}_8\}$, $\{\mathbf{A}_6, \mathbf{A}_7\}$. Since there are only two non-orthogonal dispersion matrices (modulating two real symbols) in each group, the ML decoding of **J4** can be achieved by the joint detection of two real symbols, instead of two complex symbols as reported in [8].

Table 3.1 Fulfillment of QOC for the dispersion matrices of **J4** as defined in (3.21)

(a) QOC fulfillment for **J4**

	A_1	A_2	A_3	A_4	A_5	A_6	A_7	A_8
A_1	X			X				
A_2		X	X					
A_3		X	X					
A_4	X			X				
A_5					X			X
A_6						X	X	
A_7						X	X	
A_8					X			X

(b) After re-arranging some rows and columns of (a)

	A_1	A_4	A_2	A_3	A_5	A_8	A_6	A_7
A_1	X	X						
A_4	X	X						
A_2			X	X				
A_3			X	X				
A_5					X	X		
A_8					X	X		
A_6							X	X
A_7							X	X

Legend: ☐ QO-Constraint is fulfilled
              ☒ QO-Constraint is not fulfilled

Next, Table 3.2(a) examines the fulfillment of QOC for the dispersion matrices (listed in (3.22) as matrices A_{CR_1}, A_{CR_2}, ..., A_{CR_8}) of **J4_CR**, the constellation-rotated **J4**.

$$\mathbf{A}_{CR\_1} = \begin{bmatrix} 1 & 0 & 0 & 0 \\ 0 & 1 & 0 & 0 \\ 0 & 0 & 1 & 0 \\ 0 & 0 & 0 & 1 \end{bmatrix}, \quad \mathbf{A}_{CR\_3} = \frac{1}{\sqrt{2}} \begin{bmatrix} 0 & 0 & 1+j & 0 \\ 0 & 0 & 0 & 1-j \\ -1+j & 0 & 0 & 0 \\ 0 & -1-j & 0 & 0 \end{bmatrix},$$

$$\mathbf{A}_{CR\_2} = \begin{bmatrix} 0 & 1 & 0 & 0 \\ -1 & 0 & 0 & 0 \\ 0 & 0 & 0 & 1 \\ 0 & 0 & -1 & 0 \end{bmatrix}, \quad \mathbf{A}_{CR\_4} = \frac{1}{\sqrt{2}} \begin{bmatrix} 0 & 0 & 0 & 1+j \\ 0 & 0 & -1+j & 0 \\ 0 & -1+j & 0 & 0 \\ 1+j & 0 & 0 & 0 \end{bmatrix},$$

$$\mathbf{A}_{CR\_5} = \begin{bmatrix} j & 0 & 0 & 0 \\ 0 & -j & 0 & 0 \\ 0 & 0 & -j & 0 \\ 0 & 0 & 0 & j \end{bmatrix}, \quad \mathbf{A}_{CR\_7} = \frac{j}{\sqrt{2}} \begin{bmatrix} 0 & 0 & 1+j & 0 \\ 0 & 0 & 0 & -1+j \\ 1-j & 0 & 0 & 0 \\ 0 & -1-j & 0 & 0 \end{bmatrix},$$

$$\mathbf{A}_{CR\_6} = \begin{bmatrix} 0 & j & 0 & 0 \\ j & 0 & 0 & 0 \\ 0 & 0 & 0 & -j \\ 0 & 0 & -j & 0 \end{bmatrix}, \quad \mathbf{A}_{CR\_8} = \frac{j}{\sqrt{2}} \begin{bmatrix} 0 & 0 & 0 & 1+j \\ 0 & 0 & 1-j & 0 \\ 0 & 1-j & 0 & 0 \\ 1+j & 0 & 0 & 0 \end{bmatrix}.$$

$$(3.22)$$

By re-arranging the rows and columns of Table 3.2(a) to obtain Table 3.2(b), it is clear that the dispersion matrices of **J4_CR** can only be grouped into two orthogonal groups: $\{\mathbf{A}_{CR\_1}, \mathbf{A}_{CR\_4}, \mathbf{A}_{CR\_5}, \mathbf{A}_{CR\_8}\}$ and $\{\mathbf{A}_{CR\_2}, \mathbf{A}_{CR\_3}, \mathbf{A}_{CR\_6}, \mathbf{A}_{CR\_7}\}$. Since there are four non-orthogonal dispersion matrices (modulating four real symbols) in each group, the ML decoder for **J4_CR** needs to jointly decode four real symbols, which is a two-fold increase from the two real symbols required by **J4** before CR.

Table 3.2 Fulfillment of QOC for the dispersion matrices of **J4_CR** as defined in (3.22)

(a) QO-Constraint fulfillment for **J4_CR**

	A_{CR1}	A_{CR2}	A_{CR3}	A_{CR4}	A_{CR5}	A_{CR6}	A_{CR7}	A_{CR8}
A_{CR1}	X			X				
A_{CR2}		X	X					
A_{CR3}		X	X			X	X	
A_{CR4}	X			X	X			X
A_{CR5}				X	X			X
A_{CR6}			X			X	X	
A_{CR7}			X			X	X	
A_{CR8}				X	X			X

(b) After re-arranging rows and columns of (a)

	A_{CR1}	A_{CR4}	A_{CR5}	A_{CR8}	A_{CR2}	A_{CR3}	A_{CR6}	A_{CR7}
A_{CR1}	X	X						
A_{CR4}	X	X	X	X				
A_{CR5}		X	X	X				
A_{CR8}		X	X	X				
A_{CR2}					X	X		
A_{CR3}					X	X	X	X
A_{CR6}						X	X	X
A_{CR7}						X	X	X

Legend: ☐   QO-Constraint is fulfilled
            X   QO-Constraint is not fulfilled

It can similarly be shown that for the other QO-STBCs **J8** and **TBH8**, the number of real symbols required for joint detection is doubled after CR, as summarized in Table 3.3.

Table 3.3 Number of real symbols required for joint detection for QO-STBC with square- and rectangular-QAM

Tx Antennas	Code Rate	Non full-diversity QO-STBC		Full-diversity QO-STBC with CR	
4	1	**J4**	2	**J4_CR**	4
8	3/4	**J8**	2	**J8_CR**	4
8	1	**TBH8**	4	**TBH8_CR**	8

3.4 Group-Constrained Linear Transformation

3.4.1 Definition of GCLT

In order to enable a QO-STBC to achieve full diversity and maximum coding gain while maintaining the original symbol groupings and hence decoding complexity, the *Group-Constrained Linear Transformation* (GCLT) as defined in *Proposition 3.2* may be used to optimize the code dispersion matrices.

Proposition 3.2: By linearly combining the dispersion matrices $\mathbf{A}_i$ *within the same group* in accordance with (3.23) and (3.24), the resultant dispersion matrices, $\mathbf{A}_{\mathrm{LT}\_i}$, will satisfy the QOC stated in (3.20) with the same symbol grouping structure as the original matrices $\mathbf{A}_i$.

$$\tilde{\mathbf{A}}_{\mathrm{LT}\_i} = \sum_{k \in \mathcal{G}(i)} \alpha_{i,k} \mathbf{A}_k, \quad \forall\ 1 \le i \le 2K, \tag{3.23}$$

$$\mathbf{A}_{\mathrm{LT}\_i} = \phi_i \tilde{\mathbf{A}}_{\mathrm{LT}\_i}, \quad \forall\ 1 \le i \le 2K, \tag{3.24}$$

where $\alpha_{i,p}$ are the *GCLT parameters* and are real constants, and the scaling factor

$$\phi_i = \sqrt{\frac{PN_T}{K} \Big/ \mathrm{Tr}(\tilde{\mathbf{A}}_{\mathrm{LT}\_i}^{\mathrm{H}} \tilde{\mathbf{A}}_{\mathrm{LT}\_i})} \tag{3.25}$$

serves to ensure that every dispersion matrix of the QO-STBC after GCLT have equal transmission power, i.e. satisfying the power distribution constraint in (1.15) (ii) or MSD constraint in (3.6).

Proof of Proposition 3.2:   Applying (3.20) with $\mathbf{A}_p$ replaced by $\mathbf{A}_{\mathbf{LT}\_p}$ and $\mathbf{A}_q$ replaced by $\mathbf{A}_{\mathbf{LT}\_q}$ gives

$$\mathbf{A}_{\mathbf{LT}\_p}^{\mathrm{H}}\mathbf{A}_{\mathbf{LT}\_q} + \mathbf{A}_{\mathbf{LT}\_q}^{\mathrm{H}}\mathbf{A}_{\mathbf{LT}\_p} \qquad 1\le p,q \le 2K \quad \text{and} \quad q \notin \mathcal{G}(p)$$

$$= \phi_p\left(\sum_{i\in\mathcal{G}(p)}\left(\alpha_{p,i}\mathbf{A}_i\right)\right)^{\mathrm{H}}\phi_q\left(\sum_{k\in\mathcal{G}(q)}\left(\alpha_{q,k}\mathbf{A}_k\right)\right) +$$

$$\phi_q\left(\sum_{k\in\mathcal{G}(q)}\left(\alpha_{q,k}\mathbf{A}_k\right)\right)^{\mathrm{H}}\phi_p\left(\sum_{i\in\mathcal{G}(p)}\left(\alpha_{p,i}\mathbf{A}_i\right)\right)$$

$$= \phi_p\phi_q\left(\sum_{i\in\mathcal{G}(p)}\sum_{k\in\mathcal{G}(q)}\left(\alpha_{p,i}\alpha_{q,k}\mathbf{A}_i^{\mathrm{H}}\mathbf{A}_k\right)\right) + \tag{3.26}$$

$$\phi_q\phi_p\left(\sum_{k\in\mathcal{G}(q)}\sum_{i\in\mathcal{G}(p)}\left(\alpha_{q,k}\alpha_{p,i}\mathbf{A}_k^{\mathrm{H}}\mathbf{A}_i\right)\right)$$

$$= \phi_p\phi_q\sum_{i\in\mathcal{G}(p)}\sum_{k\in\mathcal{G}(q)}\left(\alpha_{p,i}\alpha_{q,k}\underbrace{\left(\mathbf{A}_i^{\mathrm{H}}\mathbf{A}_k + \mathbf{A}_k^{\mathrm{H}}\mathbf{A}_i\right)}_{=0 \text{ as per QOC}}\right)$$

$$= 0.$$

Since the above expression is equal to zero, the dispersion matrices $\{\mathbf{A}_{\mathbf{LT}}\}$ satisfy the QOC (3.20) as long as matrices $\{\mathbf{A}\}$ do, hence *Proposition 3.2* is proven.   ∎

Here we wish to mention that to obtain more general results, the power normalization operation in (3.24) and (3.25) can be relaxed to just ensure that the total transmission power of the dispersion matrices satisfies (1.15) (i) instead of (1.15) (ii). In this case, assuming that the dispersion matrices $\mathbf{A}_k$ and $\mathbf{A}_q$ are in the same group, i.e. $\{k,q\} = \mathcal{G}(k) = \mathcal{G}(q)$, we can replace (3.24) and (3.25) with (3.27) and (3.28) as follows:

$$\mathbf{A}_{\mathbf{LT}\_k} = \phi_{kq}\tilde{\mathbf{A}}_{\mathbf{LT}\_k}, \qquad \{k,q\}=\mathcal{G}(k)=\mathcal{G}(q), \qquad (3.27)$$
$$\mathbf{A}_{\mathbf{LT}\_q} = \phi_{kq}\tilde{\mathbf{A}}_{\mathbf{LT}\_q},$$

$$\phi_{kq} = \sqrt{\frac{2PN_T}{K} \Bigg/ \left(\mathrm{Tr}\left(\tilde{\mathbf{A}}^{\mathrm{H}}_{\mathbf{LT}\_k}\tilde{\mathbf{A}}_{\mathbf{LT}\_k}\right) + \mathrm{Tr}\left(\tilde{\mathbf{A}}^{\mathrm{H}}_{\mathbf{LT}\_q}\tilde{\mathbf{A}}_{\mathbf{LT}\_q}\right)\right)}. \qquad (3.28)$$

3.4.2 *Optimization of GCLT parameters*

The GCLT parameters α in (3.23) can be chosen such that $\{\mathbf{A}_{\mathbf{LT}}\}$ satisfy certain chosen performance optimization criteria, such as the rank and determinant criteria discussed in Section 1.5. To provide a systematic way to optimize the GCLT parameters, the *Multi-dimensional Lattice Rotation* (MLR) technique in [49] may be employed. For simplicity, consider a QO-STBC with two real symbols per group such as **J4**. Assuming that the dispersion matrices $\mathbf{A}_k$ and $\mathbf{A}_q$ are in the same group, i.e. $\{k, q\} = \mathcal{G}(k) = \mathcal{G}(q)$, the GCLT of these dispersion matrices can be expressed as follows:

$$\begin{bmatrix} \tilde{\mathbf{A}}_{\mathbf{LT}\_k} \\ \tilde{\mathbf{A}}_{\mathbf{LT}\_q} \end{bmatrix} = \left(\mathbf{I}_{P\times P} \otimes \begin{bmatrix} \alpha_{k,k} & \alpha_{k,q} \\ \alpha_{q,k} & \alpha_{q,q} \end{bmatrix} \right) \begin{bmatrix} \mathbf{A}_k \\ \mathbf{A}_q \end{bmatrix} = \left(\mathbf{I}_{P\times P} \otimes \mathbf{L}_{\mathrm{MLR}} \right) \begin{bmatrix} \mathbf{A}_k \\ \mathbf{A}_q \end{bmatrix}, \quad (3.29)$$

where $\otimes$ represents the Kronecker product and $\mathbf{L}_{\mathrm{MLR}}$ is an orthogonal matrix as specified in [49]. For example, for a two-dimensional case, $\mathbf{L}_{\mathrm{MLR}}$ maps four GCLT parameters into one variable θ using:

$$\mathbf{L}_{\mathrm{MLR}} = \begin{bmatrix} \alpha_{k,k} & \alpha_{k,q} \\ \alpha_{q,k} & \alpha_{q,q} \end{bmatrix} = \begin{bmatrix} \cos(\theta) & \sin(\theta) \\ -\sin(\theta) & \cos(\theta) \end{bmatrix}. \qquad (3.30)$$

This streamlines the search or analysis of the optimum GCLT parameters because it reduces the number of optimization variables from four (i.e. $\alpha_{k,k}$, $\alpha_{k,q}$, $\alpha_{q,k}$, $\alpha_{q,q}$) to one (i.e. θ).

3.4.2.1 GCLT of **J4**

Denoting the **J4** code after GCLT as **J4_LT**, the optimization of its GCLT parameters can be derived as follows. The determinant expression for the codeword distance matrix of **J4_LT** can be shown to be:

$$\det = \left[\begin{array}{l} \left((\tilde{\Delta}_1 + \tilde{\Delta}_4)^2 + (\tilde{\Delta}_2 - \tilde{\Delta}_3)^2 + (\tilde{\Delta}_5 + \tilde{\Delta}_8)^2 + (\tilde{\Delta}_6 - \tilde{\Delta}_7)^2 \right) \times \\ \left((\tilde{\Delta}_1 - \tilde{\Delta}_4)^2 + (\tilde{\Delta}_2 + \tilde{\Delta}_3)^2 + (\tilde{\Delta}_5 - \tilde{\Delta}_8)^2 + (\tilde{\Delta}_6 + \tilde{\Delta}_7)^2 \right) \end{array} \right]^2 , \quad (3.31)$$

where $\tilde{\Delta}_k = \Delta_k \cos\theta - \Delta_k \sin\theta$ and $\tilde{\Delta}_q = \Delta_p \sin\theta + \Delta_q \cos\theta$ for $(k,q) \in \{(1,4), (2,3), (5,8), (6,7)\}$, and Δ_k represents the possible error in the real PAM symbol d_k. Note that (k, q) follows the symbol grouping in **J4** indicated in Table 3.1.

Since d_1 and d_4 are in the same group of **J4_LT**, and they are independent of (i.e. separable from) the other symbols in the ML decoding operation, the worst-case (i.e. minimum) determinant value in (3.31) can be simplified by assuming that only d_1 and d_4 have errors and the other symbols are error-free [32] (i.e. assuming that $\tilde{\Delta}_2 = \tilde{\Delta}_3 = \tilde{\Delta}_5 = \tilde{\Delta}_6 = \tilde{\Delta}_7 = \tilde{\Delta}_8 = 0$):

$$\begin{aligned} \det &= \left[(\tilde{\Delta}_1 + \tilde{\Delta}_4)^2 \times (\tilde{\Delta}_1 - \tilde{\Delta}_4)^2 \right]^2 \\ &= \left[(\tilde{\Delta}_1)^2 - (\tilde{\Delta}_4)^2 \right]^4 \\ &= \left[(\Delta_1 \cos\theta - \Delta_4 \sin\theta)^2 - (\Delta_1 \sin\theta + \Delta_4 \cos\theta)^2 \right]^4 \\ &= \left[\Delta_1^2 \cos(2\theta) - 2\Delta_1\Delta_4 \sin(2\theta) - \Delta_4^2 \cos(2\theta) \right]^4 . \end{aligned} \quad (3.32)$$

4-QAM Constellation

Consider first the optimization of GCLT parameters of **J4_LT** with 4-QAM constellation. The I and Q components of a 4-QAM symbol can be viewed as two independent 2-PAM symbols. Hence $\Delta_1, \Delta_4 \in \{0, \pm d_{min}\}$ where d_{min} is the minimum Euclidian distance between two constellation points as shown in Fig. 3.4, and Δ_1 and Δ_4 cannot be both zero in (3.32). To maximize the minimum determinant value in (3.32) based on the Rank and Determinant Criteria presented in Section 1.5, we consider the following four cases of (Δ_1, Δ_4) and their resultant determinant values in (3.32):

$$\text{Case 1: } (\Delta_1, \Delta_4) = \pm (0, d_{\min}) \rightarrow \det_1 = d_{\min}^8 \left[\cos(2\theta)\right]^4; \qquad (3.33)$$

$$\text{Case 2: } (\Delta_1, \Delta_4) = \pm (d_{\min}, 0) \rightarrow \det_2 = d_{\min}^8 \left[\cos(2\theta)\right]^4; \qquad (3.34)$$

$$\text{Case 3: } (\Delta_1, \Delta_4) = \pm (d_{\min}, d_{\min}) \rightarrow \det_3 = d_{\min}^8 \left[-2\sin(2\theta)\right]^4; \quad (3.35)$$

$$\text{Case 4: } (\Delta_1, \Delta_4) = \pm (d_{\min}, -d_{\min}) \rightarrow \det_4 = d_{\min}^8 \left[2\sin(2\theta)\right]^4. \quad (3.36)$$

Note that $\det_1 = \det_2$ and $\det_3 = \det_4$. In order to maximize the smaller value between $\det_1$ and $\det_3$, we equate $\det_1$ and $\det_3$ to get:

$$\cos(2\theta_{opt}) = 2\sin(2\theta_{opt})$$
$$\Rightarrow \tan(2\theta_{opt}) = 1/2 \qquad (3.37)$$
$$\Rightarrow \theta_{opt} = \frac{1}{2}\tan^{-1}(\frac{1}{2}) = 13.28^0.$$

The result will be likewise be obtained by considering d_2 and d_3, d_5 and d_8, d_6 and d_7 separately.

So the optimum GCLT parameters for **J4_LT** are:

$$\alpha_{k,k} = \alpha_{q,q} = \cos(\theta_{opt}) = \cos(13.28^0);$$
$$\alpha_{k,q} = -\alpha_{q,k} = \sin(\theta_{opt}) = \sin(13.28^0), \qquad (3.38)$$

for $(k,q) \in \{(1,4), (2,3), (5,8), (6,7)\}$, and the minimum determinant value of the codeword distance matrix is $0.64d_{\min}^8$. Compared with **J4_CR**, which has a minimum determinant value $d_{\min}^8$ [31], **J4_LT** has a slightly smaller minimum determinant value, but the ML decoding of **J4_LT** requires the joint detection of only half the number of symbols required by **J4_CR**, and the slight reduction in the minimum determinant value will be shown later to give only less than 0.5dB loss in coding gain.

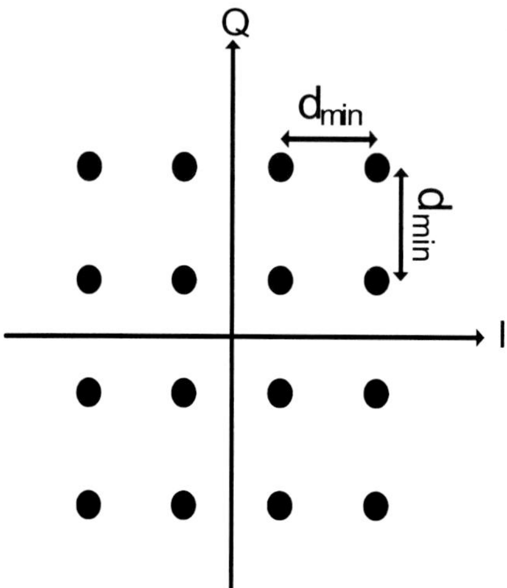

Fig. 3.4 QAM-constellation and its minimum Euclidean distance d_{min}

M-ary Square-QAM Constellation

We now derive the optimum GCLT parameters of **J4_LT** with larger QAM constellations. Consider the M-ary square-QAM constellation, where the I and Q components of a symbol can be viewed as two independent and identical $\sqrt{M}$-ary PAM symbols. The following four cases of (Δ_1, Δ_4) and their resultant determinant values as per (3.32) are considered:

Case 1: $(\Delta_1, \Delta_4) = \pm (0, nd_{min})$

$$\rightarrow \det_5 = d_{min}^8 \left[n^2 \cos(2\theta) \right]^4 ; \tag{3.39}$$

Case 2: $(\Delta_1, \Delta_4) = \pm (md_{min}, 0)$

$$\rightarrow \det_6 = d_{min}^8 \left[m^2 \cos(2\theta) \right]^4 ; \tag{3.40}$$

Case 3: $(\Delta_1, \Delta_4) = \pm (md_{min}, nd_{min})$

$$\rightarrow \det_7 = d_{min}^8 \left[(m^2 - n^2)\cos(2\theta) - 2mn\sin(2\theta) \right]^4 ; \tag{3.41}$$

Case 4: $(\Delta_1, \Delta_4) = \pm (md_{min}, -nd_{min})$

$$\rightarrow \det_8 = d_{min}^8 \left[(m^2 - n^2)\cos(2\theta) + 2mn\sin(2\theta) \right]^4 , \tag{3.42}$$

where m and n are integers such that $1 \le m, n \le \sqrt{M} - 1$, and M is the cardinality of the QAM constellation.

To maximize the smaller value of $\det_5$ to $\det_8$ for all values of m and n, consider first the smallest value of $m = n = 1$. For this case, $\det_5$ to $\det_8$ are identical to $\det_1$ to $\det_4$, hence the optimum θ value for (3.39) to (3.42) is the same as that derived in (3.37), i.e. $\theta_{\mathrm{opt}} = \frac{1}{2} \tan^{-1}(\frac{1}{2})$, and the corresponding $\det_5$ to $\det_8$ values are identically $0.64 d_{\min}^8$.

Next, consider $m, n > 1$. In this case, it can be shown that with $\theta = \frac{1}{2} \tan^{-1}(\frac{1}{2})$,

$$\det_5\big|_{m,n>1} = n^8 \left(\det_5\big|_{m=n=1} \right) \ge \det_5\big|_{m=n=1} ;$$

$$\det_6\big|_{m,n>1} = m^8 \left(\det_6\big|_{m=n=1} \right) \ge \det_6\big|_{m=n=1} ;$$

$$\det_7\big|_{m,n>1} = \left[(m^2 - mn - n^2) \right]^4 \left(\det_7\big|_{m=n=1} \right) \ge \det_7\big|_{m=n=1} ;$$

$$\det_8\big|_{m,n>1} = \left[(m^2 + mn - n^2) \right]^4 \left(\det_8\big|_{m=n=1} \right) \ge \det_8\big|_{m=n=1} ,$$

(3.43)

which are all greater than or equal to $0.64 d_{\min}^8$. Hence the worst-case (minimum) determinant value for **J4_LT** with M-ary square-QAM constellation occurs when $m = n = 1$, and it is optimized when $\theta = \theta_{\mathrm{opt}} = \frac{1}{2} \tan^{-1}(\frac{1}{2})$.

M-ary Rectangular-QAM Constellation

This is similar to the case of square-QAM, except that m and n may have different maximum values. Since it has been shown in the square-QAM case that the worst-case determinant value depends on the case of $m = n = 1$, the θ_{opt} in (3.37) and the optimum GCLT parameters in (3.38) also apply to rectangular-QAM. Therefore, we have shown that the optimum GCLT parameters derived in (3.38) apply to all sizes of square and rectangular QAM.

To verify the above result graphically, the determinant values of **J4_LT** with square 16-QAM constellation is plotted as a function of θ in Fig. 3.5. For square 16-QAM, the values of m and n can each have value 1, 2 or 3. A few combinations of m and n are considered in Fig. 3.5 for illustration purpose. The loci of the minimum determinant value over all lines are traced out in green color. It can be seen that θ_{opt} corresponds to the max-min determinant value of the code, which is in turn given by the intersection of the $\det_7|_{m=3,n=2}$, $\det_7|_{m=2,n=1}$, $\det_8|_{m=1,n=1}$, $\det_5|_{m=1,n=1}$.

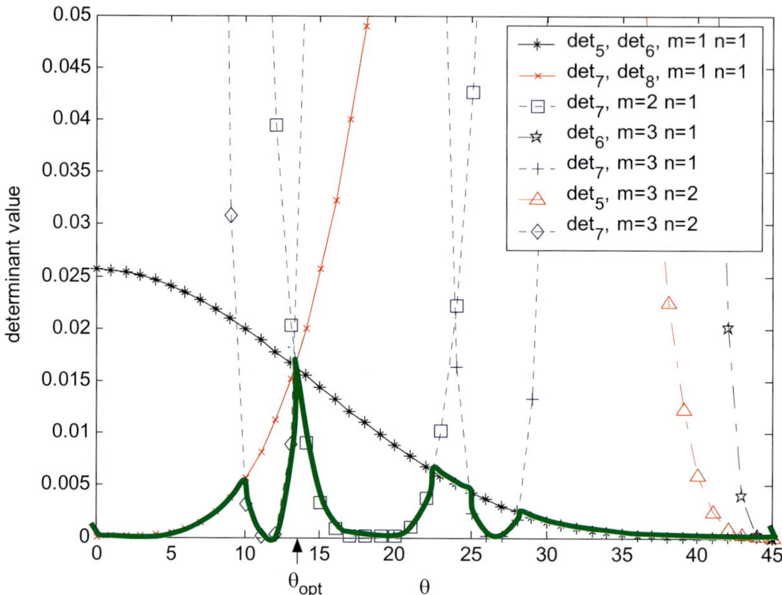

Fig. 3.5 Determinant values of **J4_LT** in (3.39) to (3.42) for various *m*, *n* values

The dispersion matrices of **J4_LT** obtained based on the optimum GCLT parameters derived in (3.38) are shown in (3.44) as $\mathbf{A}_{LT\_1}$ to $\mathbf{A}_{LT\_8}$. In Table 3.4, **J4_LT** is shown to have exactly the same number of symbol groups as **J4**, hence the GCLT technique is able to optimize the decoding performance of QO-STBC while maintaining its decoding complexity.

$$\mathbf{A}_{LT\_1} = \begin{bmatrix} a & 0 & 0 & b \\ 0 & a & -b & 0 \\ 0 & -b & a & 0 \\ b & 0 & 0 & a \end{bmatrix}, \quad \mathbf{A}_{LT\_2} = \begin{bmatrix} 0 & a & b & 0 \\ -a & 0 & 0 & b \\ -b & 0 & 0 & a \\ 0 & -b & -a & 0 \end{bmatrix},$$

$$\mathbf{A}_{LT\_3} = \begin{bmatrix} 0 & -b & a & 0 \\ b & 0 & 0 & a \\ -a & 0 & 0 & -b \\ 0 & -a & b & 0 \end{bmatrix}, \quad \mathbf{A}_{LT\_4} = \begin{bmatrix} -b & 0 & 0 & a \\ 0 & -b & -a & 0 \\ 0 & -a & -b & 0 \\ a & 0 & 0 & -b \end{bmatrix},$$

$$\mathbf{A}_{LT\_5} = \begin{bmatrix} ja & 0 & 0 & jb \\ 0 & -ja & jb & 0 \\ 0 & jb & -ja & 0 \\ jb & 0 & 0 & ja \end{bmatrix}, \quad \mathbf{A}_{LT\_6} = \begin{bmatrix} 0 & ja & jb & 0 \\ ja & 0 & 0 & -jb \\ jb & 0 & 0 & -ja \\ 0 & -jb & -ja & 0 \end{bmatrix},$$

$$\mathbf{A}_{LT\_7} = \begin{bmatrix} 0 & -jb & ja & 0 \\ -jb & 0 & 0 & -ja \\ ja & 0 & 0 & jb \\ 0 & -ja & jb & 0 \end{bmatrix}, \quad \mathbf{A}_{LT\_8} = \begin{bmatrix} -jb & 0 & 0 & ja \\ 0 & jb & ja & 0 \\ 0 & ja & jb & 0 \\ ja & 0 & 0 & -jb \end{bmatrix}, \quad (3.44)$$

where $a = \cos(13.28^0)$ and $b = \sin(13.28^0)$.

Table 3.4 Fulfillment of QOC for the dispersion matrices of **J4_LT** as defined in (3.44)

(a) QOC fulfillment for **J4_LT**

	$\mathbf{A}_{LT1}$	$\mathbf{A}_{LT2}$	$\mathbf{A}_{LT3}$	$\mathbf{A}_{LT4}$	$\mathbf{A}_{LT5}$	$\mathbf{A}_{LT6}$	$\mathbf{A}_{LT7}$	$\mathbf{A}_{LT8}$
$\mathbf{A}_{LT1}$	X			X				
$\mathbf{A}_{LT2}$		X	X					
$\mathbf{A}_{LT3}$		X	X					
$\mathbf{A}_{LT4}$	X			X				
$\mathbf{A}_{LT5}$					X			X
$\mathbf{A}_{LT6}$						X	X	
$\mathbf{A}_{LT7}$						X	X	
$\mathbf{A}_{LT8}$					X			X

(b) After re-arranging some rows and columns of (a)

	$\mathbf{A}_{LT1}$	$\mathbf{A}_{LT4}$	$\mathbf{A}_{LT2}$	$\mathbf{A}_{LT3}$	$\mathbf{A}_{LT5}$	$\mathbf{A}_{LT8}$	$\mathbf{A}_{LT6}$	$\mathbf{A}_{LT7}$
$\mathbf{A}_{LT1}$	X	X						
$\mathbf{A}_{LT4}$	X	X						
$\mathbf{A}_{LT2}$			X	X				
$\mathbf{A}_{LT3}$			X	X				
$\mathbf{A}_{LT5}$					X	X		
$\mathbf{A}_{LT8}$					X	X		
$\mathbf{A}_{LT6}$							X	X
$\mathbf{A}_{LT7}$							X	X

Legend: ☐ QO-Constraint is fulfilled
                ☒ QO-Constraint is not fulfilled

3.4.2.2   *GCLT of **J8***

Since the rate-3/4 8×8 QO-STBC **J8** is also double–symbol–decodable as **J4**, it can similarly be shown that the optimum GCLT parameters for **J8** with M-ary QAM constellation are:

$$\alpha_{k,k} = \alpha_{q,q} = \cos(13.28^0),$$
$$\alpha_{k,q} = -\alpha_{q,k} = \sin(13.28^0), \tag{3.45}$$

for $(k, q) \in \{(1, 10), (2, 11), (3, 12), (4, 7), (5, 8), (6, 9)\}$. The resultant code, denoted as **J8_LT**, has minimum determinant value $0.4096(\sqrt{4/3}d_{min})^{16}$, as compared with $(\sqrt{4/3}d_{min})^{16}$ for **J8_CR**.

3.4.2.3   *GCLT of **TBH8***

TBH8, a rate-1 8×8 QO-STBC, requires a joint detection of four real symbols. Its corresponding $\mathbf{L}_{MLR}$ matrix in (3.29) is [49]:

$$\mathbf{L}_{MLR} = \prod_{1 \leq i \leq 3,\ i+1 \leq k \leq 4} \mathbf{G}(i,k,\theta_{ik}), \tag{3.46}$$

where $\mathbf{G}(i, k, \theta_{ik})$ is a 4×4 matrix with entries at (i, i) and (k, k) equal to $\cos(\theta_{ik})$, entry at (i, k) equals to $\sin(\theta_{ik})$, and entry at (k, i) equals to $-\sin(\theta_{ik})$, ones on the remaining diagonal positions and zeros elsewhere. $\mathbf{G}(i, k, \theta_{ik})$ basically models a counter-clockwise rotation by θ degree with respect to the (i, k) plane. For example, for $i = 2$, $k = 3$, the $\mathbf{G}$ matrix becomes:

$$G(2,3,\theta_{23}) = \begin{bmatrix} 1 & 0 & 0 & 0 \\ 0 & \cos(\theta_{23}) & \sin(\theta_{23}) & 0 \\ 0 & -\sin(\theta_{23}) & \cos(\theta_{23}) & 0 \\ 0 & 0 & 0 & 1 \end{bmatrix}. \tag{3.47}$$

For **TBH8**, the GCLT parameters are related to $\mathbf{L_{MLR}}$ by the following relationship:

$$\begin{bmatrix} \alpha_{p,p} & \alpha_{p,q} & \alpha_{p,m} & \alpha_{p,n} \\ \alpha_{q,p} & \alpha_{q,q} & \alpha_{q,m} & \alpha_{q,n} \\ \alpha_{m,p} & \alpha_{m,q} & \alpha_{m,m} & \alpha_{m,n} \\ \alpha_{n,p} & \alpha_{n,q} & \alpha_{n,m} & \alpha_{n,n} \end{bmatrix} = \mathbf{L}_{MLR}, \tag{3.48}$$

where $(p, q, m, n) \in \{(1, 4, 6, 7), (2, 3, 5, 8), (9, 12, 14, 15), (10, 11, 13, 16)\}$. The mathematical derivation of the optimum GCLT parameters for **TBH8** is difficult because of the large dimension involved. Hence we rely on computer search to find the optimized solution. The best solution found is: $\theta_{12} = -45.66^0$, $\theta_{23} = 9.43^0$, $\theta_{34} = -46.11^0$, $\theta_{14} = 37.78^0$, $\theta_{13} = 9.13^0$, $\theta_{24} = 44.24^0$.

3.4.3 *Performance comparison*

3.4.3.1 *ML decoding complexity*

The ML decoding metrics of **J4_LT** are shown in (3.49). Note that they each rely on only two *real* symbols. In contrast, the decoding metrics of **J4_CR** (which are the same as the decoding metrics of **J4** in (2.23) except that the c_3 and c_4 symbols are rotated) rely on two *complex* symbols. Hence GCLT achieves a reduction of decoding complexity over CR. This complexity reduction will be increasingly significant with increasing QAM constellation size, as illustrated in Table 3.5.

$$f_1(d_1,d_4) = \sum_{k=1}^{N_R} \left[\begin{array}{c} (\sum_{i=1}^{4}|h_{i,k}|^2)(|ad_1 - bd_4|^2 + |bd_1 + ad_4|^2) + \\ 2\,\mathrm{Re}\left\{ \begin{array}{c} (ad_1 - bd_4)\alpha + (bd_1 + ad_4)\beta + \\ (ad_1 - bd_4)(bd_1 + ad_4)\gamma \end{array} \right\} \end{array} \right];$$

$$f_2(d_2,d_3) = \sum_{k=1}^{N_R} \left[\begin{array}{c} (\sum_{i=1}^{4}|h_{i,k}|^2)(|ad_2 - bd_3|^2 + |bd_2 + ad_3|^2) + \\ 2\,\mathrm{Re}\left\{ \begin{array}{c} (ad_2 - bd_3)\chi + (bd_2 + ad_3)\delta + \\ (ad_2 - bd_3)(bd_2 + ad_3)\varphi \end{array} \right\} \end{array} \right];$$

$$f_3(d_5,d_8) = \sum_{k=1}^{N_R} \left[\begin{array}{c} (\sum_{i=1}^{4}|h_{i,k}|^2)(|ad_5 - bd_8|^2 + |bd_5 + ad_8|^2) + \\ 2\,\mathrm{Re}\left\{ \begin{array}{c} j(ad_5 - bd_8)\alpha + j(bd_5 + ad_8)\beta + \\ (ad_5 - bd_8)(bd_5 + ad_8)\gamma \end{array} \right\} \end{array} \right];$$

$$f_4(d_6,d_7) = \sum_{k=1}^{N_R} \left[\begin{array}{c} (\sum_{i=1}^{4}|h_{i,k}|^2)(|ad_6 - bd_7|^2 + |bd_6 + ad_7|^2) + \\ 2\,\mathrm{Re}\left\{ \begin{array}{c} j(ad_6 - bd_7)\chi + j(bd_6 + ad_7)\delta + \\ (ad_6 - bd_7)(bd_6 + ad_7)\varphi \end{array} \right\} \end{array} \right], \quad (3.49)$$

where

$$\alpha = -h_{1,r}r_1^* - h_{2,r}^* r_2 - h_{3,r}^* r_3 - h_{4,r} r_4^*;$$

$$\beta = -h_{4,r}r_1^* + h_{3,r}^* r_2 + h_{2,r}^* r_3 - h_{1,r} r_4^*;$$

$$\gamma = 2\,\mathrm{Re}(h_{1,r}h_{4,r}^* - h_{2,r}h_{3,r}^*);$$

$$\chi = -h_{2,r}r_1^* + h_{1,r}^* r_2 - h_{4,r}^* r_3 + h_{3,r} r_4^*;$$

$$\delta = -h_{3,r}r_1^* - h_{4,r}^* r_2 + h_{1,r}^* r_3 + h_{2,r} r_4^*;$$

$$\varphi = 2\,\mathrm{Re}(-h_{1,r}h_{4,r}^* + h_{2,r}h_{3,r}^*),$$

and a = cos(13.28^0) and b = sin(13.28^0).

Table 3.5 Comparison of ML decoding search space for **J4_CR** and **J4_LT** for various constellations

	Decoding search space	
	J4_CR	**J4_LT**
QPSK	16	4
16QAM	256	16
Constellation with M points	M^2	M

3.4.3.2 Decoding performance

The diversity product (coding gain) defined in (1.23) of **J4_CR**, **J4_LT**, **J8_CR**, **J8_LT**, **TBH8_CR** and **TBH8_LT** with 4-QAM are listed in Table 3.6. It shows that the GCLT codes achieve slightly lower diversity products than the GCLT codes. However such drops in diversity product will be shown later to result in negligible loss in BER performance.

Table 3.6 Comparison of 4QAM QO-STBCs with CR and GCLT

	Diversity product
J4_CR [16]	0.3536
J4_LT	0.3344
J8_CR [16]	0.2887
J8_LT	0.2730
TBH8_CR [31]	0.2187
TBH8_LT	0.1531

The BER performance of the **J4**, **J4_CR** and **J4_LT** codes will next be compared, using the rate-1/2 O-STBC **G4C** from [6] as performance benchmark. Since **J4**, **J4_CR** and **J4_LT** are full-rate codes and **G4C** is a half-rate code, 16-QAM constellation is used for **G4C** while 4-QAM constellation is used for the QO-STBCs to set a common spectral efficiency of 2 bits/s/Hz for all codes. In Fig. 3.6, it is observed that both **J4_CR** [16] and **J4_LT** achieve full transmit diversity as they have the same BER slope as the **G4C**. **J4_CR** and **J4_LT** also have lower BER than the **G4C** because the former are full-rate codes with smaller QAM dimension and hence larger Euclidean distance. Although **J4_CR** has slightly better performance (corresponding to the larger diversity product shown in Table 3.6) than **J4_LT**, their BER performance difference is less than 0.3dB. **J4_LT**, however, has a much lower decoding complexity than **J4_CR** as shown earlier in Table 3.5.

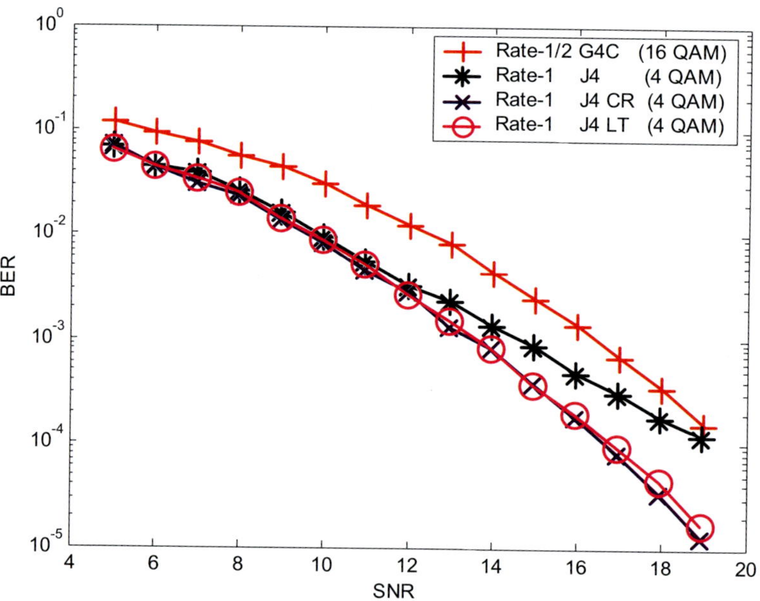

Fig. 3.6 Simulation results of QO-STBCs for four transmit antennas with spectral efficiency of 2 bits/sec/Hz

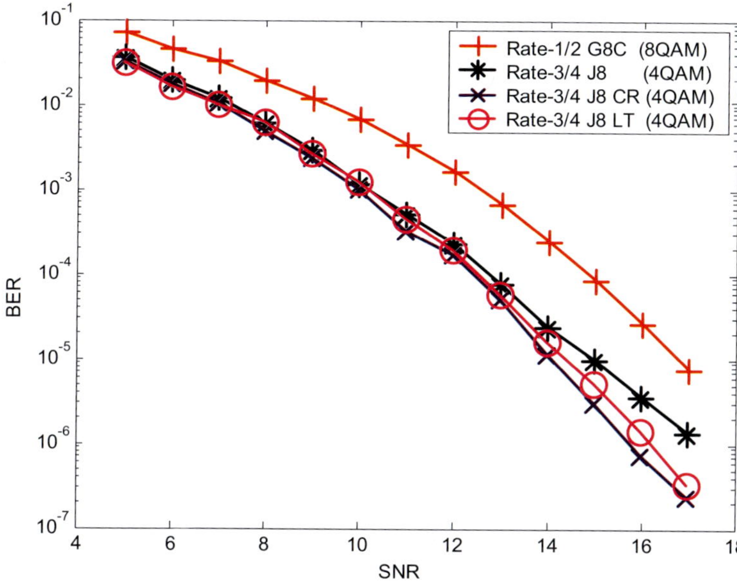

Fig. 3.7 Simulation results of QO-STBCs for eight transmit antennas with spectral efficiency of 1.5 bits/sec/Hz

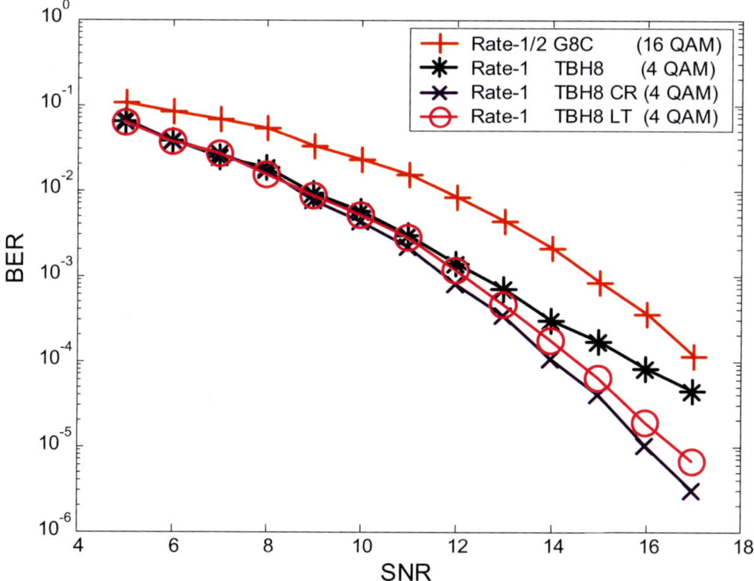

Fig. 3.8 Simulation results of QO-STBCs for eight transmit antennas with spectral efficiency of 2 bits/sec/Hz

Similar conclusions can be drawn for **J8**, **J8_CR** and **J8_LT**, as shown in Fig. 3.7, as well as for **TBH8_CR** and **TBH8_LT** as shown in Fig. 3.8. Their performances are benchmarked against **G8C**, a rate-1/2 O-STBC from [6].

3.5 Chapter Summary

In this chapter, we first derive the Quasi-Orthogonality Constraints (QOC) which unifies the algebraic structure of all existing QO-STBCs. We then whiten the received QO-STBC signal after matched filtering, so that we can consider standard decoders for QO-STBC and avoid tedious derivation of dedicated decoders. Based on the derived algebraic structure, we show that the commonly used CR technique actually increases the decoding search space of the resultant QO-STBC when regular square and rectangular QAM constellations are used. Specifically, the number of symbols required for joint detection is doubled after CR is applied. Hence we introduce GCLT as a means to optimize the decoding performance of QO-STBC without increasing the joint detection search space. The optimum GCLT parameters for achieving full diversity and coding gains are obtained for square- or rectangular-QAM constellations. Simulation results show that QO-STBC with GCLT can achieve full diversity with negligible loss in BER performance compared to QO-STBC with CR. But the decoding search space of QO-STBC with GCLT is the square root of the decoding search space of QO-STBC with CR, hence GCLT lead to a great reduction in decoding complexity over CR, especially when the QAM size is large. Although only **J4**, **J8** and **TBH8** are used as examples, the GCLT concept and its optimization methodology described in this chapter can be used to achieve the same reduction in decoding complexity for other QO-STBCs reported in [9,10,42,50,51,52] too.

Chapter 4

Quasi-Orthogonal Space-Time Block Code with Minimum Decoding Complexity

As mentioned earlier, Orthogonal Space-Time Block Code (O-STBC) is able to provide full transmit diversity with linear maximum-likelihood (ML) decoding complexity. Specifically, O-STBC with phase shift keying (PSK) constellation is single–symbol–decodable, while O-STBC with square or rectangular quadrature amplitude modulation (QAM) constellation can even be half–symbol–decodable (i.e. the real and imaginary parts of a symbol can be decoded separately). However, O-STBC suffers code rate less than one if more than two transmit antennas and complex constellation are used.

Quasi-Orthogonal Space-Time Block Code (QO-STBC) achieves a higher code rate than O-STBC at the expense of higher decoding complexity. QO-STBC also achieves full diversity when combined with constellation rotation (CR) or group-constrained linear transformation (GCLT). With CR, rate-1 QO-STBC for four transmit antennas and rate-3/4 QO-STBC for eight transmit antennas are double–symbol–decodable and require joint detection of two complex symbols per group; while rate-1 QO-STBC for eight transmit antennas is quad–symbol–decodable and requires joint detection of four complex symbols per group. With GCLT, the number of symbols required for joint detection are halved. However, GCLT is only applicable to square- or rectangular- QAM constellations.

A summary of the code rate and decoding complexity of the above-mentioned STBCs are shown in Table 4.1.

73

Table 4.1 Comparisons of STBCs for four transmit antennas

STBCs for four transmit antennas		No. of real symbols required for ML joint detection	
	Code Rate	PSK Constellation	Square- and Rectangular-QAM Constellation
O-STBC	3/4	2	1
QO-STBC with CR	1	4	4
QO-STBC with GCLT	1	4	2

Table 4.2 Comparisons of STBCs for eight transmit antennas

STBCs for eight transmit antennas		No. of real symbols required for ML joint detection	
	Code Rate	PSK Constellation	Square- and Rectangular-QAM Constellation
O-STBC	1/2	2	1
QO-STBC with CR	3/4	4	4
QO-STBC with GCLT	3/4	4	2

In this chapter, we introduce a new class of QO-STBC which is single–symbol–decodable (similar to O-STBC with PSK constellation) and requires the joint detection of only two *real* symbols, or one complex symbol equivalently. This is the lowest possible decoding complexity for any non-orthogonal STBC. Any code simpler will be an O-STBC. Hence we call it *Minimum–Decoding–Complexity QO-STBC* (MDC-QOSTBC). We shall derive its algebraic structure, introduce systematic methods for its construction, and investigate its achievable code rate. We will also compare its decoding performance, power distribution (which is related to the number of antennas to be turned off regularly) and antenna scalability (which is related to supporting a different number of transmit antennas) properties with the existing QO-STBCs.

4.1 Algebraic Structure of MDC-QOSTBC

Based on the QO-STBC concept discussed in Chapter 3, an MDC-QOSTBC which carries K complex symbols per codeword is a QO-STBC with $G = K$ groups, each containing two real symbols or equivalently one complex symbol. At the receiver, every complex symbol in an MDC-QOSTBC can be separated from all the other complex symbols by simple linear matched filtering, but the I and Q components of every complex symbol need to be jointly detected, hence MDC-QOSTBC is said to be single-symbol decodable. The joint detection of different complex symbols can be carried out separately and in parallel. This greatly reduces the decoding complexity of MDC-QOSTBC.

Definition 4.1: A *Minimum–Decoding–Complexity QO-STBC* (MDC-QOSTBC) is a QO-STBC whose equivalent channel matrix $\mathbf{H}$ as defined in (1.13) has the property that $\mathbf{H}^T\mathbf{H}$ is block-diagonal with non-zero sub-matrices of size 2×2.

Definition 4.1 is the limiting case of *Definition 3.1*. Clearly, a QO-STBC with $\mathbf{H}^T\mathbf{H}$ block-diagonal and containing sub-matrix size of 2×2 implies minimum decoding complexity for the QO-STBC, because any further reduction in the sub-matrix size leads to an O-STBC with a diagonal $\mathbf{H}^T\mathbf{H}$.

Theorem 4.1: For different complex symbols (indexed using subscripts i and k) in an MDC-QOSTBC to be separable from each other, their corresponding dispersion matrices $\{\mathbf{A}_i, \mathbf{B}_i\}$ and $\{\mathbf{A}_k, \mathbf{B}_k\}$ must possess the following algebraic structure, herein referred to as *Minimum–Decoding–Complexity Quasi-Orthogonality Constraints* (MDC-QOC):

$$\text{(i)} \quad \mathbf{A}_i^H\mathbf{A}_k = -\mathbf{A}_k^H\mathbf{A}_i;$$
$$\text{(ii)} \quad \mathbf{B}_i^H\mathbf{B}_k = -\mathbf{B}_k^H\mathbf{B}_i; \qquad 1 \le i \ne k \le K. \qquad (4.1)$$
$$\text{(iii)} \quad \mathbf{A}_i^H\mathbf{B}_k = \mathbf{B}_k^H\mathbf{A}_i,$$

Proof of Theorem 4.1: This can be easily derived from *Theorem 3.1* by having only one complex symbol in a group, i.e. $\mathcal{G}(i) = \{i\}$. ■

Note that although the properties of O-STBC in (2.2) and those of MDC-QOC in (4.1) look alike, they are actually different because (2.2) (iii) holds for all i and k, whereas (4.1) (iii) holds only for $i \neq k$.

To facilitate the discussion in the next section, we give an example in Fig. 4.1 to illustrate the grouping concept of the dispersion matrices in an MDC-QOSTBC with $K = 4$. Fig. 4.1 shows that the dispersion matrices of the MDC-QOSTBC are divided into $K = 4$ groups, with every group consisting of only two dispersion matrices, $\mathbf{A}_i$ and $\mathbf{B}_i$, of the single complex symbol c_i.

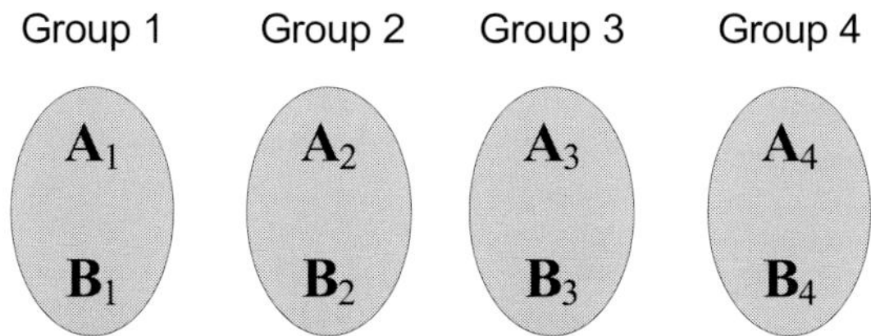

Fig. 4.1 Grouping of dispersion matrices of an MDC-QOSTBC with $K=4$

4.2 Square MDC-QOSTBC Design

Since a square STBC has minimal decoding delay according to (1.17), we will first focus on square MDC-QOSTBC design. We will compare the algebraic structures of MDC-QOSTBC and AOD, and link them using a new concept called Preferred Amicable-Orthogonal-Design Pair (Preferred AOD Pair). By doing so, we will be able to construct MDC-QOSTBC from AOD and derive the theoretical lower bound on the achievable code rate of square MDC-QOSTBC.

4.2.1 Definition of preferred AOD pair

A Preferred AOD Pair is defined as follows:

Definition 4.2: Let $\underline{\mathcal{A}}_{\mathrm{I}}$, $\underline{\mathcal{A}}_{\mathrm{II}}$, $\underline{\mathcal{B}}_{\mathrm{I}}$ and $\underline{\mathcal{B}}_{\mathrm{II}}$ be OD (Orthogonal Design, as defined in *Definition 2.2*) of the same order n where $\underline{\mathcal{A}}_{\mathrm{I}}$ is of type $(u_{\mathrm{I}\_1}, \ldots, u_{\mathrm{I}\_s_1})$ on the variables $\{\underline{a}_{\mathrm{I}\_1}, \ldots, \underline{a}_{\mathrm{I}\_s_1}\}$; $\underline{\mathcal{A}}_{\mathrm{II}}$ is of type $(u_{\mathrm{II}\_1}, \ldots, u_{\mathrm{II}\_s_2})$ on the variables $\{\underline{a}_{\mathrm{II}\_1}, \ldots, \underline{a}_{\mathrm{II}\_s_2}\}$; $\underline{\mathcal{B}}_{\mathrm{I}}$ is of type $(v_{\mathrm{I}\_1}, \ldots, v_{\mathrm{I}\_t_1})$ on the variables $\{\underline{b}_{\mathrm{I}\_1}, \ldots, \underline{b}_{\mathrm{I}\_t_1}\}$ and $\underline{\mathcal{B}}_{\mathrm{II}}$ is of type $(v_{\mathrm{II}\_1}, \ldots, v_{\mathrm{II}\_t_2})$ on the variables $\{\underline{b}_{\mathrm{II}\_1}, \ldots, \underline{b}_{\mathrm{II}\_t_2}\}$. Let $\underline{\mathcal{A}}_{\mathrm{I}}$ and $\underline{\mathcal{B}}_{\mathrm{I}}$ form an AOD and $\underline{\mathcal{A}}_{\mathrm{II}}$ and $\underline{\mathcal{B}}_{\mathrm{II}}$ form another AOD according to *Definition 2.3*. $\{\underline{\mathcal{A}}_{\mathrm{I}}, \underline{\mathcal{B}}_{\mathrm{I}}\}$ and $\{\underline{\mathcal{A}}_{\mathrm{II}}, \underline{\mathcal{B}}_{\mathrm{II}}\}$ are said to be a Preferred AOD Pair if:

1. $s_1 = t_2$,
2. $s_2 = t_1$, and
3. by shifting any one variable $\underline{\mathbf{A}}_{\mathrm{I}\_p}\,\underline{a}_{\mathrm{I}\_p}$ ($1 \le p \le s_1$) from $\underline{\mathcal{A}}_{\mathrm{I}}$ to $\underline{\mathcal{A}}_{\mathrm{II}}$ (thus resulting in a new OD $\hat{\underline{\mathcal{A}}}_{\mathrm{I}}$ with $s_1 - 1$ variables and another new OD $\hat{\underline{\mathcal{A}}}_{\mathrm{II}}$ with $s_2 + 1$ variables as shown in (4.2)), and shifting another variable $\underline{\mathbf{B}}_{\mathrm{II}\_q}\,\underline{b}_{\mathrm{II}\_q}$ ($1 \le q \le t_2$) from $\underline{\mathcal{B}}_{\mathrm{II}}$ to $\underline{\mathcal{B}}_{\mathrm{I}}$ (thus resulting in a new OD $\hat{\underline{\mathcal{B}}}_{\mathrm{II}}$ with $t_2 - 1$ variables and another new OD $\hat{\underline{\mathcal{B}}}_{\mathrm{I}}$ with $t_1 + 1$ variables as shown in (4.2)), the new pair of ODs $\{\hat{\underline{\mathcal{A}}}_{\mathrm{I}}, \hat{\underline{\mathcal{B}}}_{\mathrm{I}}\}$ forms another AOD, and so does the new pair ODs $\{\hat{\underline{\mathcal{A}}}_{\mathrm{II}}, \hat{\underline{\mathcal{B}}}_{\mathrm{II}}\}$.

$$\hat{\underline{\mathcal{A}}}_{\mathrm{I}} = \sum_{i=1}^{p-1}\underline{\mathbf{A}}_{\mathrm{I}\_i}\,\underline{a}_{\mathrm{I}\_i} + \sum_{i=p+1}^{s_1}\underline{\mathbf{A}}_{\mathrm{I}\_i}\,\underline{a}_{\mathrm{I}\_i}; \qquad \left[s_1 - 1 \text{ variables}\right]$$

$$\hat{\underline{\mathcal{B}}}_{\mathrm{I}} = \sum_{i=1}^{t_1}\underline{\mathbf{B}}_{\mathrm{I}\_i}\,\underline{b}_{\mathrm{I}\_i} + \underline{\mathbf{B}}_{\mathrm{II}\_q}\,\underline{b}_{\mathrm{II}\_q}; \qquad \left[t_1 + 1 \text{ variables}\right]$$

$$\hat{\underline{\mathcal{A}}}_{\mathrm{II}} = \sum_{i=1}^{s_2}\underline{\mathbf{A}}_{\mathrm{II}\_i}\,\underline{a}_{\mathrm{II}\_i} + \underline{\mathbf{A}}_{\mathrm{I}\_p}\,\underline{a}_{\mathrm{I}\_p}; \qquad \left[s_2 + 1 \text{ variables}\right]$$

$$\hat{\underline{\mathcal{B}}}_{\mathrm{II}} = \sum_{i=1}^{q-1}\underline{\mathbf{B}}_{\mathrm{II}\_i}\,\underline{b}_{\mathrm{II}\_i} + \sum_{i=q+1}^{t_2}\underline{\mathbf{B}}_{\mathrm{II}\_i}\,\underline{b}_{\mathrm{II}\_i}. \qquad \left[t_2 - 1 \text{ variables}\right]$$

(4.2)

The above definition of Preferred AOD Pair includes the corresponding case of shifting the variables from $\underline{\mathcal{A}}_{\mathrm{II}}$ to $\underline{\mathcal{A}}_{\mathrm{I}}$ and from $\underline{\mathcal{B}}_{\mathrm{I}}$ to $\underline{\mathcal{B}}_{\mathrm{II}}$. The new AODs formed in this case are shown in (4.3).

$$\hat{\underline{A}}_{I} = \sum_{i=1}^{s_1} \mathbf{A}_{I\_i}\, \underline{a}_{I\_i} \; + \; \mathbf{A}_{II\_p}\, \underline{a}_{II\_p}; \qquad\qquad \left[s_1 + 1 \text{ variables}\right]$$

$$\hat{\underline{B}}_{I} = \sum_{i=1}^{q-1} \mathbf{B}_{I\_i}\, \underline{b}_{I\_i} \; + \; \sum_{i=q+1}^{t_1} \mathbf{B}_{I\_i}\, \underline{b}_{I\_i}; \qquad \left[t_1 - 1 \text{ variables}\right]$$

$$\hat{\underline{A}}_{II} = \sum_{i=1}^{p-1} \mathbf{A}_{II\_i}\, \underline{a}_{II\_i} \; + \; \sum_{i=p+1}^{s_2} \mathbf{A}_{II\_i}\, \underline{a}_{II\_i}; \qquad \left[s_2 - 1 \text{ variables}\right]$$

$$\hat{\underline{B}}_{II} = \sum_{i=1}^{t_2} \mathbf{B}_{II\_i}\, \underline{b}_{II\_i} \; + \; \mathbf{B}_{I\_q}\, \underline{b}_{I\_q}. \qquad\qquad \left[t_2 + 1 \text{ variables}\right]$$

$$(4.3)$$

Proposition 4.1: The two new AODs generated by shifting a variable from a Preferred AOD Pair will themselves form a Preferred Pair.

4.2.2 Relationship between MDC-QOSTBC and AOD through preferred AOD pair

Comparing the MDC-QOC in (4.1) with the AOD Constraints in (2.11), some similarity between them can be observed:

1. The MDC-QOC (4.1) (i) and (ii) are exactly the same as the AOD Constraint (2.11) (ii). Hence they have the same solutions.
2. The MDC-QOC (4.1) (iii) is equivalent to the AOD Constraint (2.11) (iii) except that (2.11) (iii) applies for all i and k values, while (4.1) (iii) applies only for $i \neq k$. Hence solutions to (4.1) (iii) will also satisfy (2.11) (iii), but not vice versa.

In addition, the AOD Constraint (2.11) (i) is also exactly the MSD constraint in (3.6), hence MDC-QOSTBC constructed from AOD will also satisfy the MSD constraint.

The above observations suggest that there exist some relationship between the MDC-QOSTBC dispersion matrices and the AOD matrices. This relationship is illustrated in Fig. 4.2. Fig. 4.2(a) and (b) show the same eight dispersion matrices of the MDC-QOSTBC shown earlier in Fig. 4.1, except that the $\mathbf{A}_i$ and $\mathbf{B}_i$ matrix positions in the third and fourth groups in Fig. 4.2(a) are vertically swapped, while the $\mathbf{A}_i$ and $\mathbf{B}_i$ matrix positions in the first, third and fourth groups in Fig. 4.2(b) are vertically swapped (however this does not affect the group structure and decoding complexity of the MDC-QOSTBC).

First consider Fig. 4.2(a): The top row of the MDC-QOSTBC dispersion matrices $\{\mathbf{A}_1, \mathbf{A}_2\,;\, \mathbf{B}_3, \mathbf{B}_4\}$ satisfy (4.1). So they also satisfy (2.11) and they form an AOD according to Observation (2) discussed above. Similarly, the bottom row of the MDC-QOSTBC dispersion matrices $\{\mathbf{A}_3, \mathbf{A}_4\,;\, \mathbf{B}_1, \mathbf{B}_2\}$ form another AOD.

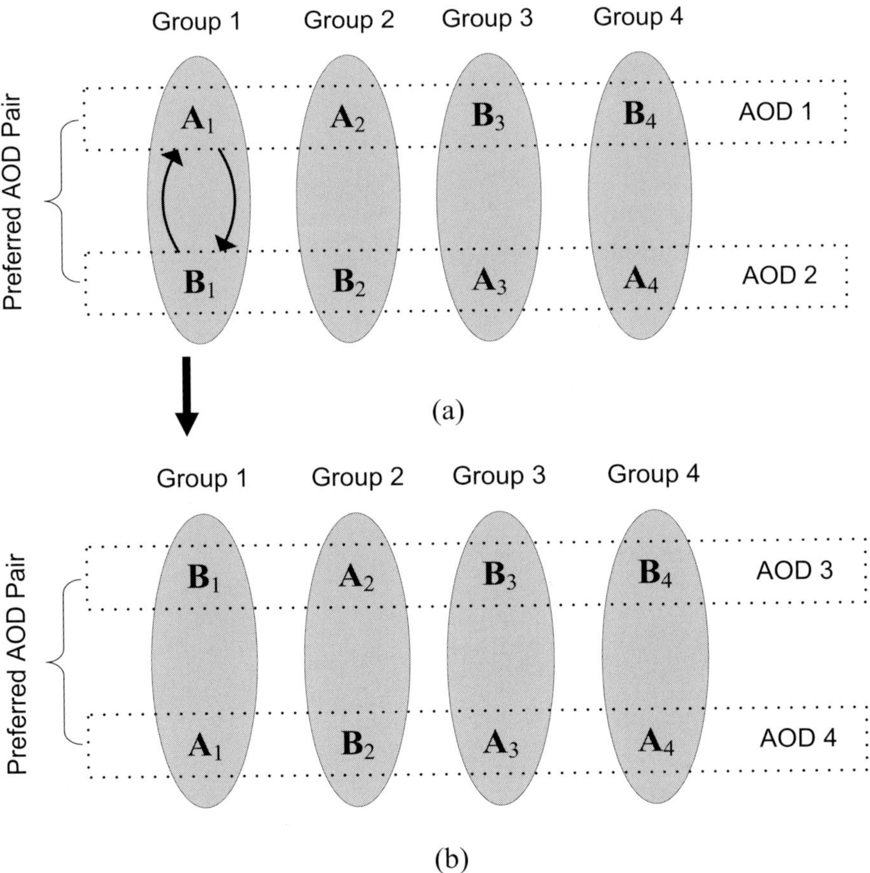

Fig. 4.2 Relationship between MDC-QOSTBC and AOD through Preferred AOD Pair

However, not any pair of AODs can form an MDC-QOSTBC because an MDC-QOSTBC requires $\{\mathbf{A}_i, \mathbf{B}_i\}$ and $\{\mathbf{A}_k, \mathbf{B}_k\}$ to satisfy (4.1) for all values of $1 \leq i \neq k \leq K$. For example, it requires $\mathbf{A}_1$ not only to satisfy (4.1) with $\mathbf{A}_2$, $\mathbf{B}_3$, and $\mathbf{B}_4$ in the top AOD, but also with $\mathbf{A}_3$, $\mathbf{A}_4$, and $\mathbf{B}_2$ in the bottom AOD in Fig. 4.2(a). So if the $\mathbf{A}_1$ and $\mathbf{B}_1$ matrices in Fig. 4.2(a) are swapped to obtain Fig. 4.2(b), and the new rows of dispersion matrices $\{\mathbf{A}_2 ; \mathbf{B}_1, \mathbf{B}_3, \mathbf{B}_4\}$ and $\{\mathbf{A}_1, \mathbf{A}_3, \mathbf{A}_4 ; \mathbf{B}_2\}$ in Fig. 4.2(b) again form two new AODs, this matches the definition of Preferred AOD Pair in *Definition 4.2*. Therefore, we have shown that an MDC-QOSTBC can be constructed by using two AODs that form a Preferred AOD Pair.

The above explanations imply that an MDC-QOSTBC for N_T transmit antennas can be constructed from two AODs of order $n = N_T$ that form a Preferred AOD Pair. Since AODs are of square designs, the code length P of the constructed square MDC-QOSTBC will be $P = n = N_T$. The number of complex symbols K transmitted by the MDC-QOSTBC is equal to the number of pairs of its dispersion matrix $\{\mathbf{A}, \mathbf{B}\}$, which is in turn given by the total number of variables of the Preferred AOD Pair divided by two (since one variable in an AOD is related to one *real* symbol, while the code rate R of STBC is typically defined as the number of *complex* symbols transmitted per unit time). Therefore, for an MDC-QOSTBC constructed from Preferred AOD Pair, its K and R can be expressed as:

$$K = \frac{s_1 + t_1 + s_2 + t_2}{2}$$

$$R = \frac{K}{P} = \frac{s_1 + t_1 + s_2 + t_2}{2n} \tag{4.4}$$

where s_1, s_2, t_1, t_2 and n are AOD variables defined in *Definition 4.2*.

4.2.3 *Lower bound on the code rate for square design*

Since (4.4) relates the code rate (R) of an MDC-QOSTBC to the number of variables (s_1, s_2, t_1, t_2) of a Preferred AOD Pair, it provides an avenue for us to derive the achievable code rate of square MDC-QOSTBC, which is an important property for any QO-STBC. To achieve this goal, we need first to determine the maximum number of variables that a Preferred AOD Pair can support. Note that since the AOD Constraints in (2.11) is stricter than the MDC-QOC in (4.1) (AOD has an extra requirement of (2.11) (i)), this exercise will lead to a lower bound on the achievable code rate of square MDC-QOSTBC.

Following the framework in [21] in deriving the upper bounds of the s and t values of AOD (i.e. first fix the value of t, then find the maximum permissible value of s), we present six propositions to derive the upper and lower bounds for the s_1, s_2, t_1 and t_2 values of a Preferred AOD Pair [53].

First consider the case that all $\mathcal{A}_I$, $\mathcal{A}_{II}$, $\mathcal{B}_I$ and $\mathcal{B}_{II}$ have non-zero number of variables.

Proposition 4.2: The maximum number of variables, t_1 and t_2, that $\mathcal{B}_I$ and $\mathcal{B}_{II}$ as defined in *Definition 4.2* can have respectively is $\min\{\rho(n)-1,\hat{t}\}$, where $\hat{t}$ is a value such that $\rho_{\hat{t}-1}(n) \geq 1$.

Proof of Proposition 4.2: According to the definition of Preferred AOD Pair in *Definition 4.2*, the OD $\mathcal{B}_I$ may be given one variable from $\mathcal{B}_{II}$ to form a new OD $\hat{\mathcal{B}}_I$ with t_1+1 variables as shown in (4.2). Since $\hat{\mathcal{B}}_I$ is an OD whose maximum number of variables must be less than $\rho(n)$ as shown in (2.12), we can conclude that $t_1+1 \leq \rho(n)$, or $t_1 \leq \rho(n)-1$. In addition, referring to (4.3), we can also take one variable from $\mathcal{B}_I$ and give it to $\mathcal{B}_{II}$. In the mean time $\mathcal{A}_I$ must accept one variable from $\mathcal{A}_{II}$ to form $\hat{\mathcal{A}}_I$. In this case, $\hat{\mathcal{A}}_I$ must have at least one variable, hence it is required that an $\hat{\mathcal{A}}_I$ with at least one variable exists, i.e. $\rho_{t_1-1}(n) \geq 1$ by (2.12). Since $\rho_t(n)$ is a decreasing function of t, t_1 should be smaller than a value $\hat{t}$ where $\rho_{\hat{t}-1}(n) \geq 1$. Combining all of the above, the maximum value of t_1 should be the smaller value of $\rho(n)-1$ or $\hat{t}$ where $\rho_{\hat{t}-1}(n) \geq 1$. Similar argument applies to t_2. Hence *Proposition 4.2* is proved. ∎

Proposition 4.3: The maximum number of variables, s_1, of $\underline{\mathcal{A}}_{\mathbf{I}}$ is $\min\{\rho_{t_1+1}(n)+1, \rho_{t_1}(n), \rho_{t_1-1}(n)-1\}$. Similarly, the maximum number of variables, s_2, of $\underline{\mathcal{A}}_{\mathbf{II}}$ is $\min\{\rho_{t_2+1}(n)+1, \rho_{t_2}(n), \rho_{t_2-1}(n)-1\}$.

Proof of Proposition 4.3: Considering (4.2), if there are t_1+1 variables in $\hat{\mathcal{B}}_{\mathbf{I}}$, there will be s_1-1 variables in $\hat{\mathcal{A}}_{\mathbf{I}}$. Similarly, if there are t_1-1 variables in $\hat{\mathcal{B}}_{\mathbf{I}}$ in (4.3), there will be s_1+1 variables in $\hat{\mathcal{A}}_{\mathbf{I}}$. Hence for an AOD to exist, by (2.12) we have $s_1-1 \leq \rho_{t_1+1}(n)$, or $s_1 \leq \rho_{t_1+1}(n)+1$; and $s_1+1 \leq \rho_{t_1-1}(n)$, or $s_1 \leq \rho_{t_1-1}(n) -1$. In addition, since $\hat{\mathcal{A}}_{\mathbf{I}}$ and $\hat{\mathcal{B}}_{\mathbf{I}}$ are derived respectively from $\underline{\mathcal{A}}_{\mathbf{I}}$ and $\mathcal{B}_{\mathbf{I}}$ that are themselves an AOD, (2.12) implies that $s_1 \leq \rho_{t_1}(n)$. Combining all the above, s_1 must satisfy all the three constraints, i.e. $s_1 \leq \min\{\rho_{t_1+1}(n)+1, \rho_{t_1}(n), \rho_{t_1-1}(n)-1\}$. Similar proof applies to s_2. $\blacksquare$

So far we have considered non-zero values for s_1, s_2, t_1 and t_2. We now consider the remaining cases of $s_2 = t_1 = 0$ or $s_1 = t_2 = 0$. First consider the case of $s_2 = t_1 = 0$, which corresponds to $\underline{\mathcal{A}}_{\mathbf{I}}$ shifting a variable to $\underline{\mathcal{A}}_{\mathbf{II}}$ and $\underline{\mathcal{B}}_{\mathbf{II}}$ shifting a variable to $\underline{\mathcal{B}}_{\mathbf{I}}$. In this case, the upper bound for s_1 and t_2 are obtained as follows:

Proposition 4.4: The maximum number of variables, t_2, that $\underline{\mathcal{B}}_{\mathbf{II}}$ can have is $\min\{\rho(n), \hat{t}\}$, where $\hat{t}$ is a value such that $\rho_{\hat{t}-1}(n) \geq 1$.

Proof of Proposition 4.4: Since $\underline{\mathcal{B}}_{\mathbf{II}}$ is an OD, its maximum number of variable, t_2, must satisfy $t_2 \leq \rho(n)$ according to (2.12). Furthermore, since $\underline{\mathcal{A}}_{\mathbf{II}}$ will accept one variable from $\underline{\mathcal{A}}_{\mathbf{I}}$ to form $\hat{\mathcal{A}}_{\mathbf{II}}$, $\hat{\mathcal{A}}_{\mathbf{II}}$ must have at least one variable. This implies that an $\hat{\mathcal{A}}_{\mathbf{II}}$ with at least one variable must exist, i.e. $\rho_{t_2-1}(n) \geq 1$ by (2.12). Since $\rho_t(n)$ is a decreasing function of t, t_2 should be smaller than a value $\hat{t}$ where $\rho_{\hat{t}-1}(n) \geq 1$. Combining all of the above, the maximum value of t_2 should be the smaller value of $\rho(n)$ or $\hat{t}$ where $\rho_{\hat{t}-1}(n) \geq 1$. As a result *Proposition 4.4* is proved. $\blacksquare$

Proposition 4.5: The maximum number of variables, s_1, that $\underline{\mathcal{A}}_\mathbf{I}$ can have is $\min\{\rho_1(n)+1, \rho_0(n)\}$.

Proof of Proposition 4.5: Since $\underline{\mathcal{B}}_\mathbf{I}$ has $t_1 = 0$, s_1 must satisfy $s_1 \leq \rho_0(n)$ according to (2.12). After shifting one variable from $\underline{\mathcal{A}}_\mathbf{I}$ to $\underline{\mathcal{A}}_\mathbf{II}$ and from $\underline{\mathcal{B}}_\mathbf{II}$ to $\underline{\mathcal{B}}_\mathbf{I}$, $\hat{\mathcal{A}}_\mathbf{I}$ will have s_1-1 variables and $\hat{\mathcal{B}}_\mathbf{I}$ will have one variable. In order for $\hat{\mathcal{A}}_\mathbf{I}$ and $\hat{\mathcal{B}}_\mathbf{I}$ to form an AOD, we must have $s_1-1 \leq \rho_1(n) \Rightarrow s_1 \leq \rho_1(n)+1$. Hence *Proposition 4.5* is proved. ∎

Finally, consider the case of $s_1 = t_2 = 0$, which corresponds to $\underline{\mathcal{A}}_\mathbf{II}$ shifting a variable to $\underline{\mathcal{A}}_\mathbf{I}$ and $\underline{\mathcal{B}}_\mathbf{I}$ shifting a variable to $\underline{\mathcal{B}}_\mathbf{II}$. In this case, the upper bound for s_2 and t_1 can be obtained as follows:

Proposition 4.6: The maximum number of variables, t_1, that $\underline{\mathcal{B}}_\mathbf{I}$ can have is $\min\{\rho(n), \hat{t}\}$, where $\hat{t}$ is a value such that $\rho_{\hat{t}-1}(n) \geq 1$.
Proof of Proposition 4.6: Similar to *Proposition 4.4*. ∎

Proposition 4.7: The maximum number of variables, s_2, that $\underline{\mathcal{A}}_\mathbf{II}$ can have is $\min\{\rho_1(n)+1, \rho_0(n)\}$.
Proof of Proposition 4.7: Similar to *Proposition 4.5*. ∎

From the above six propositions, the upper bounds for s_1, s_2, t_1 and t_2 can be summarized as follows:

when s_1, t_1, s_2, $t_2 \neq 0$:

$$t_1, t_2 \leq \min\{\rho(n)-1, \hat{t}\}, \text{ where } \hat{t} \text{ is a value such that } \rho_{\hat{t}-1}(n) \geq 1.$$
$$s_1 \leq \min\{\rho_{t_1+1}(n)+1, \rho_{t_1}(n), \rho_{t_1-1}(n)-1\} \qquad \text{and}$$
$$s_2 \leq \min\{\rho_{t_2+1}(n)+1, \rho_{t_2}(n), \rho_{t_2-1}(n)-1\}.$$

when $s_2 = t_1 = 0$:

$$t_2 \leq \min\{\rho(n), \hat{t}\}, \text{ where } \hat{t} \text{ is a value such that } \rho_{\hat{t}-1}(n) \geq 1.$$
$$s_1 \leq \min\{\rho_1(n)+1, \rho_0(n)\}.$$

when $s_1=t_2=0$:

$$t_1 \leq \min\{\rho(n),\hat{\imath}\}, \text{ where } \hat{\imath} \text{ is a value such that } \rho_{\hat{\imath}-1}(n)\geq 1.$$
$$s_2 \leq \min\{\rho_1(n)+1,\rho_0(n)\}$$

Considering the case of MDC-QOSTBC with 4 and 8 transmit antennas, we now use the results developed above to list the upper bounds of s_1, s_2, t_1 and t_2 for $n = 4$ and $n = 8$ in Table 4.3 and Table 4.4 respectively.

Table 4.3 *s, t* values for $n = 4$

t_1	$s_1 \leq$	t_2	$s_2 \leq$	Obtained from	Max($s_1+ t_1+ s_2+ t_2$)
4	0	0	4	*Proposition 4.6* *Proposition 4.7*	8
3	1	1	3	*Proposition 4.2* *Proposition 4.3*	8
2	2	2	2		8
1	3	3	1		8
0	4	4	0	*Proposition 4.4* *Proposition 4.5*	8

Table 4.4 *s, t* values for $n = 8$

t_1	$s_1 \leq$	t_2	$s_2 \leq$	Obtained from	Max($s_1+ t_1+ s_2+ t_2$)
6	0	0	6	*Proposition 4.6* *Proposition 4.7*	12
5	1	1	5		12
4	2	2	4		12
3	3	3	3	*Proposition 4.2* *Proposition 4.3*	12
2	4	4	2		12
1	5	5	1		12
0	6	6	0	*Proposition 4.4* *Proposition 4.5*	12

Interestingly, we note that the maximum value, max($s_1+t_1+s_2+t_2$) for order 4 in Table 4.3 is always a constant value of 8, while the maximum value, max($s_1+t_1+s_2+t_2$) for order 8 in Table 4.4 is always a constant value of 12.

Theorem 4.2: The lower bound on the achievable code rate of square MDC-QOSTBC is 1 for four transmit antennas and 3/4 for eight transmit antennas.

Proof of Theorem 4.2: Since the maximum total number of variables of a Preferred AOD Pair of order 4 is 8, it follows from (4.4) that the achievable code rate of an MDC-QOSTBC for four transmit antennas is lower bounded by $8/(2\times4)=1$. Likewise, for an MDC-QOSTBC for eight transmit antennas, which can be constructed by a Preferred AOD Pair of order 8 with maximum total number of variables of 12, the achievable code rate is lower bounded by $12/(2\times8)=3/4$. ∎

4.2.4 Construction of preferred AOD pair

4.2.4.1 Quaternion

Quaternion can be represented using the following complex 2×2 matrices [54]:

$$\mathbb{U} \equiv \begin{bmatrix} 1 & 0 \\ 0 & 1 \end{bmatrix}; \quad \mathbb{I} \equiv \begin{bmatrix} j & 0 \\ 0 & -j \end{bmatrix}; \quad \mathbb{J} \equiv \begin{bmatrix} 0 & 1 \\ -1 & 0 \end{bmatrix}; \quad \mathbb{K} \equiv \begin{bmatrix} 0 & j \\ j & 0 \end{bmatrix}. \quad (4.5)$$

These matrices are closely related to the Pauli matrices [54]. They possess properties that will be used later in this chapter for constructing MDC-QOSTBC from AOD.

4.2.4.2 Systematic construction of preferred AOD pair

In this section, we provide a generalized scheme to construct a Preferred AOD Pair (with complex and real entries) from lower order AODs.

Theorem 4.3: Consider an AOD $\{\underset{\sim}{\mathbf{A}};\mathbf{B}\}$ of order n, a Preferred AOD Pair (with complex and real entries) of order $2n$ can be constructed by:

$$\begin{cases} \underline{\mathbf{A}}_{\mathrm{I}\_i} = \underset{\sim}{\mathbf{A}}_i \otimes \mathbf{M}; \\ \underline{\mathbf{B}}_{\mathrm{I}\_i} = \underset{\sim}{\mathbf{B}}_i \otimes \mathbf{M}; \end{cases}$$

$$\begin{cases} \underline{\mathbf{A}}_{\mathrm{II}\_i} = \underset{\sim}{\mathbf{B}}_i \otimes j\mathbf{N}; \\ \underline{\mathbf{B}}_{\mathrm{II}\_i} = \underset{\sim}{\mathbf{A}}_i \otimes j\mathbf{N}, \end{cases} \tag{4.6}$$

where $\otimes$ represents Kronecker product and $\mathbf{M}$ and $\mathbf{N}$ are arbitrary matrices satisfying

$$\begin{aligned} &\text{(i) } \mathbf{M}^{\mathrm{H}}\mathbf{M} = \mathbf{N}^{\mathrm{H}}\mathbf{N} = \mathbf{I}, \\ &\text{(ii) } \mathbf{M}^{\mathrm{H}}\mathbf{N} = \mathbf{N}^{\mathrm{H}}\mathbf{M}. \end{aligned} \tag{4.7}$$

Proof of Theorem 4.3: First we verify that $\underline{\mathbf{A}}_{\mathrm{I}\_i}$ and $\underline{\mathbf{A}}_{\mathrm{II}\_i}$ satisfy the MDC-QOC in (4.1):

$$\begin{aligned} &\left(\underline{\mathbf{A}}_{\mathrm{I}\_i}\right)^{\mathrm{H}}\left(\underline{\mathbf{A}}_{\mathrm{II}\_i}\right) + \left(\underline{\mathbf{A}}_{\mathrm{II}\_i}\right)^{\mathrm{H}}\left(\underline{\mathbf{A}}_{\mathrm{I}\_i}\right) \\ &= \left(\underset{\sim}{\mathbf{A}}_i \otimes \mathbf{M}\right)^{\mathrm{H}}\left(\underset{\sim}{\mathbf{B}}_i \otimes j\mathbf{N}\right) + \left(\underset{\sim}{\mathbf{B}}_i \otimes j\mathbf{N}\right)^{\mathrm{H}}\left(\underset{\sim}{\mathbf{A}}_i \otimes \mathbf{M}\right) \\ &= \left[\underset{\sim}{\mathbf{A}}_i^{\mathrm{H}}\underset{\sim}{\mathbf{B}}_i \otimes \left(j\mathbf{M}^{\mathrm{H}}\mathbf{N}\right)\right] + \left[\underset{\sim}{\mathbf{B}}_i^{\mathrm{H}}\underset{\sim}{\mathbf{A}}_i \otimes \left(-j\mathbf{N}^{\mathrm{H}}\mathbf{M}\right)\right] \\ &= j\left[\left(\underset{\sim}{\mathbf{A}}_i^{\mathrm{H}}\underset{\sim}{\mathbf{B}}_i \otimes \mathbf{M}^{\mathrm{H}}\mathbf{N}\right) - \left(\underset{\sim}{\mathbf{A}}_i^{\mathrm{H}}\underset{\sim}{\mathbf{B}}_i \otimes \mathbf{M}^{\mathrm{H}}\mathbf{N}\right)^{\mathrm{H}}\right] = \mathbf{0}. \end{aligned} \tag{4.8}$$

In (4.9), we have used the relationships $\underset{\sim}{\mathbf{A}}_i^{\mathrm{H}}\underset{\sim}{\mathbf{B}}_i = \underset{\sim}{\mathbf{B}}_i^{\mathrm{H}}\underset{\sim}{\mathbf{A}}_i$ from the property of AOD in (2.11)(iii), $\mathbf{M}^{\mathrm{H}}\mathbf{N} = \mathbf{N}^{\mathrm{H}}\mathbf{M}$ from (4.7) (ii), and the property of the Kronecker product from (A.3) (iii).

Following the same steps as (4.8), we can also verify that $\underline{\mathbf{A}}_{\mathrm{I}\_i}$ and $\underline{\mathbf{B}}_{\mathrm{II}\_k}$ $(k \neq i)$ satisfy the MDC-QOC in (4.1):

$$\begin{aligned} &\left(\underline{\mathbf{A}}_{\mathrm{I}\_i}\right)^{\mathrm{H}}\left(\underline{\mathbf{B}}_{\mathrm{II}\_k}\right) - \left(\underline{\mathbf{B}}_{\mathrm{II}\_k}\right)^{\mathrm{H}}\left(\underline{\mathbf{A}}_{\mathrm{I}\_i}\right) \qquad i \neq k \\ &= \left(\underset{\sim}{\mathbf{A}}_i \otimes \mathbf{M}\right)^{\mathrm{H}}\left(\underset{\sim}{\mathbf{A}}_k \otimes j\mathbf{N}\right) - \left(\underset{\sim}{\mathbf{A}}_k \otimes j\mathbf{N}\right)^{\mathrm{H}}\left(\underset{\sim}{\mathbf{A}}_i \otimes \mathbf{M}\right) \\ &= \left[\underset{\sim}{\mathbf{A}}_i^{\mathrm{H}}\underset{\sim}{\mathbf{A}}_k \otimes \left(j\mathbf{M}^{\mathrm{H}}\mathbf{N}\right)\right] - \left[\underset{\sim}{\mathbf{A}}_k^{\mathrm{H}}\underset{\sim}{\mathbf{A}}_i \otimes \left(-j\mathbf{N}^{\mathrm{H}}\mathbf{M}\right)\right] \\ &= j\left[\left(\underset{\sim}{\mathbf{A}}_i^{\mathrm{H}}\underset{\sim}{\mathbf{A}}_k \otimes \mathbf{M}^{\mathrm{H}}\mathbf{N}\right) + \left(\underset{\sim}{\mathbf{A}}_k^{\mathrm{H}}\underset{\sim}{\mathbf{A}}_i \otimes \mathbf{M}^{\mathrm{H}}\mathbf{N}\right)^{\mathrm{H}}\right] = \mathbf{0}. \end{aligned} \tag{4.9}$$

The proofs of MDC-QOC compliance between $\underline{\mathbf{B}}_{\mathrm{I}\_i}$ and $\underline{\mathbf{B}}_{\mathrm{II}\_i}$, $\underline{\mathbf{B}}_{\mathrm{I}\_i}$ and $\underline{\mathbf{A}}_{\mathrm{II}\_k}$ etc. are similar to the above.   ■

An example of $\{\mathbf{M}, \mathbf{N}\}$ satisfying the constraints in (4.7) is:

$$\mathbf{M} \in \pm\{\mathbb{U}, \mathbb{I}\},$$
$$\mathbf{N} \in \pm j \times \{\mathbb{J}, \mathbb{K}\},$$

$$(4.10)$$

where the $\mathbb{U}$, $\mathbb{I}$, $\mathbb{J}$, $\mathbb{K}$ are the Quaternion matrices from (4.5).

The Preferred AOD Pair constructed using (4.10) and *Theorem 4.3* may consist of complex entries. To obtain Preferred AOD Pair with only real entries, $\mathbf{M} = \mathbb{U}$ and $\mathbf{N} = \mathbb{J}$ should be used, and the AOD matrices $\{\underline{\mathbf{A}}; \underline{\mathbf{B}}\}$ should contain only real entries. One such example of a Preferred AOD Pair constructed using

$$\{\underline{\mathbf{A}}; \underline{\mathbf{B}}\} = \left\{ \begin{bmatrix} 1 & 0 \\ 0 & 1 \end{bmatrix}, \begin{bmatrix} 0 & 1 \\ -1 & 0 \end{bmatrix}; \begin{bmatrix} 1 & 0 \\ 0 & -1 \end{bmatrix}, \begin{bmatrix} 0 & 1 \\ 1 & 0 \end{bmatrix} \right\}$$

$$(4.11)$$

is shown in (4.12) and (4.13).

Amicable Orthogonal Design I of order 4:

$$\underline{\mathbf{A}}_{I\_1} = \begin{bmatrix} 1 & 0 & 0 & 0 \\ 0 & 1 & 0 & 0 \\ 0 & 0 & 1 & 0 \\ 0 & 0 & 0 & 1 \end{bmatrix}, \quad \underline{\mathbf{A}}_{I\_2} = \begin{bmatrix} 0 & 1 & 0 & 0 \\ -1 & 0 & 0 & 0 \\ 0 & 0 & 0 & 1 \\ 0 & 0 & -1 & 0 \end{bmatrix},$$

$$\underline{\mathbf{B}}_{I\_1} = \begin{bmatrix} 1 & 0 & 0 & 0 \\ 0 & -1 & 0 & 0 \\ 0 & 0 & 1 & 0 \\ 0 & 0 & 0 & -1 \end{bmatrix}, \quad \underline{\mathbf{B}}_{I\_2} = \begin{bmatrix} 0 & 1 & 0 & 0 \\ 1 & 0 & 0 & 0 \\ 0 & 0 & 0 & 1 \\ 0 & 0 & 1 & 0 \end{bmatrix}.$$

$$(4.12)$$

Amicable Orthogonal Design II of order 4:

$$\mathbf{A}_{\text{II\_1}} = \begin{bmatrix} 0 & 0 & -1 & 0 \\ 0 & 0 & 0 & 1 \\ 1 & 0 & 0 & 0 \\ 0 & -1 & 0 & 0 \end{bmatrix}, \quad \mathbf{A}_{\text{II\_2}} = \begin{bmatrix} 0 & 0 & 0 & -1 \\ 0 & 0 & -1 & 0 \\ 0 & 1 & 0 & 0 \\ 1 & 0 & 0 & 0 \end{bmatrix},$$

$$\mathbf{B}_{\text{II\_1}} = \begin{bmatrix} 0 & 0 & -1 & 0 \\ 0 & 0 & 0 & -1 \\ 1 & 0 & 0 & 0 \\ 0 & 1 & 0 & 0 \end{bmatrix}, \quad \mathbf{B}_{\text{II\_2}} = \begin{bmatrix} 0 & 0 & 0 & -1 \\ 0 & 0 & 1 & 0 \\ 0 & 1 & 0 & 0 \\ -1 & 0 & 0 & 0 \end{bmatrix}. \tag{4.13}$$

Another example of real Preferred AOD Pair of order 8 constructed using

$$\{\mathbf{A};\mathbf{B}\} = \left\{ \begin{matrix} \begin{bmatrix} 1 & 0 & 0 & 0 \\ 0 & 1 & 0 & 0 \\ 0 & 0 & 1 & 0 \\ 0 & 0 & 0 & 1 \end{bmatrix} \begin{bmatrix} 0 & 0 & 1 & 0 \\ 0 & 0 & 0 & 1 \\ -1 & 0 & 0 & 0 \\ 0 & -1 & 0 & 0 \end{bmatrix} \begin{bmatrix} 0 & 0 & 0 & -1 \\ 0 & 0 & 1 & 0 \\ 0 & -1 & 0 & 0 \\ 1 & 0 & 0 & 0 \end{bmatrix}; \\[6pt] \begin{bmatrix} 1 & 0 & 0 & 0 \\ 0 & 1 & 0 & 0 \\ 0 & 0 & -1 & 0 \\ 0 & 0 & 0 & -1 \end{bmatrix} \begin{bmatrix} 0 & 0 & 1 & 0 \\ 0 & 0 & 0 & -1 \\ 1 & 0 & 0 & 0 \\ 0 & -1 & 0 & 0 \end{bmatrix} \begin{bmatrix} 0 & 0 & 0 & 1 \\ 0 & 0 & 1 & 0 \\ 0 & 1 & 0 & 0 \\ 1 & 0 & 0 & 0 \end{bmatrix} \end{matrix} \right\} \tag{4.14}$$

is shown in (4.15) and (4.16).

Amicable Orthogonal Design I of order 8:

$$\underline{\mathbf{A}}_{I\_1} = \begin{bmatrix} 1 & 0 & 0 & 0 & 0 & 0 & 0 & 0 \\ 0 & 1 & 0 & 0 & 0 & 0 & 0 & 0 \\ 0 & 0 & 1 & 0 & 0 & 0 & 0 & 0 \\ 0 & 0 & 0 & 1 & 0 & 0 & 0 & 0 \\ 0 & 0 & 0 & 0 & 1 & 0 & 0 & 0 \\ 0 & 0 & 0 & 0 & 0 & 1 & 0 & 0 \\ 0 & 0 & 0 & 0 & 0 & 0 & 1 & 0 \\ 0 & 0 & 0 & 0 & 0 & 0 & 0 & 1 \end{bmatrix}, \quad \underline{\mathbf{A}}_{I\_2} = \begin{bmatrix} 0 & 0 & 1 & 0 & 0 & 0 & 0 & 0 \\ 0 & 0 & 0 & 1 & 0 & 0 & 0 & 0 \\ -1 & 0 & 0 & 0 & 0 & 0 & 0 & 0 \\ 0 & -1 & 0 & 0 & 0 & 0 & 0 & 0 \\ 0 & 0 & 0 & 0 & 0 & 0 & 1 & 0 \\ 0 & 0 & 0 & 0 & 0 & 0 & 0 & 1 \\ 0 & 0 & 0 & 0 & -1 & 0 & 0 & 0 \\ 0 & 0 & 0 & 0 & 0 & -1 & 0 & 0 \end{bmatrix},$$

$$\underline{\mathbf{A}}_{I\_3} = \begin{bmatrix} 0 & 0 & 0 & -1 & 0 & 0 & 0 & 0 \\ 0 & 0 & 1 & 0 & 0 & 0 & 0 & 0 \\ 0 & -1 & 0 & 0 & 0 & 0 & 0 & 0 \\ 1 & 0 & 0 & 0 & 0 & 0 & 0 & 0 \\ 0 & 0 & 0 & 0 & 0 & 0 & 0 & -1 \\ 0 & 0 & 0 & 0 & 0 & 0 & 1 & 0 \\ 0 & 0 & 0 & 0 & 0 & -1 & 0 & 0 \\ 0 & 0 & 0 & 0 & 1 & 0 & 0 & 0 \end{bmatrix}, \quad \underline{\mathbf{B}}_{I\_1} = \begin{bmatrix} 1 & 0 & 0 & 0 & 0 & 0 & 0 & 0 \\ 0 & 1 & 0 & 0 & 0 & 0 & 0 & 0 \\ 0 & 0 & -1 & 0 & 0 & 0 & 0 & 0 \\ 0 & 0 & 0 & -1 & 0 & 0 & 0 & 0 \\ 0 & 0 & 0 & 0 & 1 & 0 & 0 & 0 \\ 0 & 0 & 0 & 0 & 0 & 1 & 0 & 0 \\ 0 & 0 & 0 & 0 & 0 & 0 & -1 & 0 \\ 0 & 0 & 0 & 0 & 0 & 0 & 0 & -1 \end{bmatrix},$$

$$\underline{\mathbf{B}}_{I\_2} = \begin{bmatrix} 0 & 0 & 1 & 0 & 0 & 0 & 0 & 0 \\ 0 & 0 & 0 & -1 & 0 & 0 & 0 & 0 \\ 1 & 0 & 0 & 0 & 0 & 0 & 0 & 0 \\ 0 & -1 & 0 & 0 & 0 & 0 & 0 & 0 \\ 0 & 0 & 0 & 0 & 0 & 0 & 1 & 0 \\ 0 & 0 & 0 & 0 & 0 & 0 & 0 & -1 \\ 0 & 0 & 0 & 0 & 1 & 0 & 0 & 0 \\ 0 & 0 & 0 & 0 & 0 & -1 & 0 & 0 \end{bmatrix}, \quad \underline{\mathbf{B}}_{I\_3} = \begin{bmatrix} 0 & 0 & 0 & 1 & 0 & 0 & 0 & 0 \\ 0 & 0 & 1 & 0 & 0 & 0 & 0 & 0 \\ 0 & 1 & 0 & 0 & 0 & 0 & 0 & 0 \\ 1 & 0 & 0 & 0 & 0 & 0 & 0 & 0 \\ 0 & 0 & 0 & 0 & 0 & 0 & 0 & 1 \\ 0 & 0 & 0 & 0 & 0 & 0 & 1 & 0 \\ 0 & 0 & 0 & 0 & 0 & 1 & 0 & 0 \\ 0 & 0 & 0 & 0 & 1 & 0 & 0 & 0 \end{bmatrix}.$$

$$(4.15)$$

Amicable Orthogonal Design II of order 8:

$$\underline{\mathbf{A}}_{\text{II\_1}} = \begin{bmatrix} 0 & 0 & 0 & 0 & -1 & 0 & 0 & 0 \\ 0 & 0 & 0 & 0 & 0 & -1 & 0 & 0 \\ 0 & 0 & 0 & 0 & 0 & 0 & 1 & 0 \\ 0 & 0 & 0 & 0 & 0 & 0 & 0 & 1 \\ 1 & 0 & 0 & 0 & 0 & 0 & 0 & 0 \\ 0 & 1 & 0 & 0 & 0 & 0 & 0 & 0 \\ 0 & 0 & -1 & 0 & 0 & 0 & 0 & 0 \\ 0 & 0 & 0 & -1 & 0 & 0 & 0 & 0 \end{bmatrix}, \quad \underline{\mathbf{A}}_{\text{II\_2}} = \begin{bmatrix} 0 & 0 & 0 & 0 & 0 & 0 & -1 & 0 \\ 0 & 0 & 0 & 0 & 0 & 0 & 0 & 1 \\ 0 & 0 & 0 & 0 & -1 & 0 & 0 & 0 \\ 0 & 0 & 0 & 0 & 0 & 1 & 0 & 0 \\ 0 & 0 & 1 & 0 & 0 & 0 & 0 & 0 \\ 0 & 0 & 0 & -1 & 0 & 0 & 0 & 0 \\ 1 & 0 & 0 & 0 & 0 & 0 & 0 & 0 \\ 0 & -1 & 0 & 0 & 0 & 0 & 0 & 0 \end{bmatrix},$$

$$\underline{\mathbf{A}}_{\text{II\_3}} = \begin{bmatrix} 0 & 0 & 0 & 0 & 0 & 0 & 0 & -1 \\ 0 & 0 & 0 & 0 & 0 & 0 & -1 & 0 \\ 0 & 0 & 0 & 0 & 0 & -1 & 0 & 0 \\ 0 & 0 & 0 & 0 & -1 & 0 & 0 & 0 \\ 0 & 0 & 0 & 1 & 0 & 0 & 0 & 0 \\ 0 & 0 & 1 & 0 & 0 & 0 & 0 & 0 \\ 0 & 1 & 0 & 0 & 0 & 0 & 0 & 0 \\ 1 & 0 & 0 & 0 & 0 & 0 & 0 & 0 \end{bmatrix}, \quad \underline{\mathbf{B}}_{\text{II\_1}} = \begin{bmatrix} 0 & 0 & 0 & 0 & -1 & 0 & 0 & 0 \\ 0 & 0 & 0 & 0 & 0 & -1 & 0 & 0 \\ 0 & 0 & 0 & 0 & 0 & 0 & -1 & 0 \\ 0 & 0 & 0 & 0 & 0 & 0 & 0 & -1 \\ 1 & 0 & 0 & 0 & 0 & 0 & 0 & 0 \\ 0 & 1 & 0 & 0 & 0 & 0 & 0 & 0 \\ 0 & 0 & 1 & 0 & 0 & 0 & 0 & 0 \\ 0 & 0 & 0 & 1 & 0 & 0 & 0 & 0 \end{bmatrix},$$

$$\underline{\mathbf{B}}_{\text{II\_2}} = \begin{bmatrix} 0 & 0 & 0 & 0 & 0 & 0 & -1 & 0 \\ 0 & 0 & 0 & 0 & 0 & 0 & 0 & -1 \\ 0 & 0 & 0 & 0 & 1 & 0 & 0 & 0 \\ 0 & 0 & 0 & 0 & 0 & 1 & 0 & 0 \\ 0 & 0 & 1 & 0 & 0 & 0 & 0 & 0 \\ 0 & 0 & 0 & 1 & 0 & 0 & 0 & 0 \\ -1 & 0 & 0 & 0 & 0 & 0 & 0 & 0 \\ 0 & -1 & 0 & 0 & 0 & 0 & 0 & 0 \end{bmatrix}, \quad \underline{\mathbf{B}}_{\text{II\_3}} = \begin{bmatrix} 0 & 0 & 0 & 0 & 0 & 0 & 0 & 1 \\ 0 & 0 & 0 & 0 & 0 & 0 & -1 & 0 \\ 0 & 0 & 0 & 0 & 0 & 1 & 0 & 0 \\ 0 & 0 & 0 & 0 & -1 & 0 & 0 & 0 \\ 0 & 0 & 0 & -1 & 0 & 0 & 0 & 0 \\ 0 & 0 & 1 & 0 & 0 & 0 & 0 & 0 \\ 0 & -1 & 0 & 0 & 0 & 0 & 0 & 0 \\ 1 & 0 & 0 & 0 & 0 & 0 & 0 & 0 \end{bmatrix}.$$

$$(4.16)$$

4.2.4.3 Examples of MDC-QOSTBC constructed from preferred AOD pair

In the Preferred AOD Pair (4.12) and (4.13), $\{\underline{\mathbf{A}}_{\mathrm{I}\_1}, \underline{\mathbf{A}}_{\mathrm{I}\_2}; \underline{\mathbf{B}}_{\mathrm{I}\_1}, \underline{\mathbf{B}}_{\mathrm{I}\_2}\}$ form an AOD of order 4, similarly $\{\underline{\mathbf{A}}_{\mathrm{II}\_1}, \underline{\mathbf{A}}_{\mathrm{II}\_2}; \underline{\mathbf{B}}_{\mathrm{II}\_1}, \underline{\mathbf{B}}_{\mathrm{II}\_2}\}$ form another AOD of order 4. These two AODs form a Preferred AOD Pair with $s_1 = t_1 = s_2 = t_2 = 2$, they achieve the upper bound of s and t shown in the shaded row in Table 4.3. Consider shifting $\underline{\mathbf{A}}_{\mathrm{I}\_1}$ from the former AOD to the latter AOD, and $\underline{\mathbf{B}}_{\mathrm{II}\_1}$ from the latter AOD to the former AOD, two new AODs of order 4, $\{\underline{\mathbf{A}}_{\mathrm{I}\_2}; \underline{\mathbf{B}}_{\mathrm{I}\_1}, \underline{\mathbf{B}}_{\mathrm{I}\_2}, \underline{\mathbf{B}}_{\mathrm{II}\_1}\}$ and $\{\underline{\mathbf{A}}_{\mathrm{I}\_1}, \underline{\mathbf{A}}_{\mathrm{II}\_1}, \underline{\mathbf{A}}_{\mathrm{II}\_2}; \underline{\mathbf{B}}_{\mathrm{II}\_2}\}$, will be formed. These two new AODs also form a Preferred AOD Pair with $s_1 = t_2 = 1$ and $t_1 = s_2 = 3$, which achieves the upper bound of variables in the third row of Table 4.3. By continuing to interchange the remaining pairs of $\underline{\mathbf{A}}$ and $\underline{\mathbf{B}}$ matrices between the two AODs, Preferred AOD Pairs that achieve all the other upper bound of variables listed in Table 4.3 can be obtained. This shows that Preferred AOD Pair exists, and the upper bound on its number of variables can be reached.

Using the Preferred AOD Pair of order 4 in (4.12) and (4.13), a new full-rate MDC-QOSTBC for four transmit antenna as shown in (4.17) is formed using the dispersion matrix assignments in (4.18).

$$\begin{bmatrix} c_1^R + jc_3^I & c_2^R + jc_4^I & -c_3^R - jc_1^I & -c_4^R - jc_2^I \\ -c_2^R + jc_4^I & c_1^R - jc_3^I & -c_4^R + jc_2^I & c_3^R - jc_1^I \\ c_3^R + jc_1^I & c_4^R + jc_2^I & c_1^R + jc_3^I & c_2^R + jc_4^I \\ c_4^R - jc_2^I & -c_3^R + jc_1^I & -c_2^R + jc_4^I & c_1^R - jc_3^I \end{bmatrix}. \tag{4.17}$$

$$\begin{aligned} \mathbf{A}_1 &= \underline{\mathbf{A}}_{\mathrm{I}\_1}; & \mathbf{A}_2 &= \underline{\mathbf{A}}_{\mathrm{I}\_2}; & \mathbf{A}_3 &= \underline{\mathbf{A}}_{\mathrm{II}\_1}; & \mathbf{A}_4 &= \underline{\mathbf{A}}_{\mathrm{II}\_2}; \\ \mathbf{B}_1 &= \underline{\mathbf{B}}_{\mathrm{II}\_1}; & \mathbf{B}_2 &= \underline{\mathbf{B}}_{\mathrm{II}\_2}; & \mathbf{B}_3 &= \underline{\mathbf{B}}_{\mathrm{I}\_1}; & \mathbf{B}_4 &= \underline{\mathbf{B}}_{\mathrm{I}\_2}. \end{aligned} \tag{4.18}$$

Likewise, the Preferred AOD Pair of order 8 with $s_1 = t_1 = s_2 = t_2 = 3$ in (4.15) and (4.16) can be shown to achieve all the upper bounds of variables listed in Table 4.4. They can be used to construct the rate-3/4 MDC-QOSTBC for eight transmit antenna in (4.19) using the dispersion matrix assignments in (4.20).

$$
\begin{bmatrix}
c_1^R + jc_4^I & 0 & c_2^R + jc_5^I & -c_3^R + jc_6^I & -c_4^R - jc_1^I & 0 & -c_5^R - jc_2^I & -c_6^R + jc_3^I \\
0 & c_1^R + jc_4^I & c_3^R + jc_6^I & c_2^R - jc_5^I & 0 & -c_4^R - jc_1^I & -c_6^R - jc_3^I & c_5^R - jc_2^I \\
-c_2^R + jc_5^I & -c_3^R + jc_6^I & c_1^R - jc_4^I & 0 & -c_5^R + jc_2^I & -c_6^R + jc_3^I & c_4^R - jc_1^I & 0 \\
c_3^R + jc_6^I & -c_2^R - jc_5^I & 0 & c_1^R - jc_4^I & -c_6^R - jc_3^I & c_5^R + jc_2^I & 0 & c_4^R - jc_1^I \\
c_4^R + jc_1^I & 0 & c_5^R + jc_2^I & c_6^R - jc_3^I & c_1^R + jc_4^I & 0 & c_2^R + jc_5^I & -c_3^R + jc_6^I \\
0 & c_4^R + jc_1^I & c_6^R + jc_3^I & -c_5^R + jc_2^I & 0 & c_1^R + jc_4^I & c_3^R + jc_6^I & c_2^R - jc_5^I \\
c_5^R - jc_2^I & c_6^R - jc_3^I & -c_4^R + jc_1^I & 0 & -c_2^R + jc_5^I & -c_3^R + jc_6^I & c_1^R - jc_4^I & 0 \\
c_6^R + jc_3^I & -c_5^R - jc_2^I & 0 & -c_4^R + jc_1^I & c_3^R + jc_6^I & -c_2^R - jc_5^I & 0 & c_1^R - jc_4^I
\end{bmatrix}.
$$

$$
\tag{4.19}
$$

$$
\mathbf{A}_1 = \underline{\mathbf{A}}_{I\_1}; \quad \mathbf{A}_2 = \underline{\mathbf{A}}_{I\_2}; \quad \mathbf{A}_3 = \underline{\mathbf{A}}_{I\_3}; \quad \mathbf{A}_4 = \underline{\mathbf{A}}_{II\_1}; \quad \mathbf{A}_5 = \underline{\mathbf{A}}_{II\_2}; \quad \mathbf{A}_6 = \underline{\mathbf{A}}_{II\_3};
$$

$$
\mathbf{B}_1 = \underline{\mathbf{B}}_{II\_1}; \quad \mathbf{B}_2 = \underline{\mathbf{B}}_{II\_2}; \quad \mathbf{B}_3 = \underline{\mathbf{B}}_{II\_3}; \quad \mathbf{B}_4 = \underline{\mathbf{B}}_{I\_1}; \quad \mathbf{B}_5 = \underline{\mathbf{B}}_{I\_2}; \quad \mathbf{B}_6 = \underline{\mathbf{B}}_{I\_3}.
$$

$$
\tag{4.20}
$$

Note that *Theorem 4.3* is not the only way to construct Preferred AOD Pair and MDC-QOSTBC. This is because (4.6) and (4.7) are sufficient conditions for constructing a Preferred AOD Pair, but not necessary conditions. In the next section, we will present another MDC-QOSTBC construction which does not come under *Theorem 4.3*.

4.3   Construction of MDC-QOSTBC from O-STBC

In Section 4.2, we investigated the algebraic relationship between MDC-QOSTBC and AOD, and showed that MDC-QOSTBC can be constructed using a Preferred AOD Pair of order n equal to the number of transmit antenna. Such MDC-QOSTBC construction is restricted to the square design due to the fact that AOD is square. In this section, we present another method to construct MDC-QOSTBC, this time from a lower-order O-STBC [55,56]. This new approach has the benefit of relying on many well-established results on O-STBC, and it can be extended to non-square MDC-QOSTBC design.

4.3.1 Construction method

Construction 4.1: Consider an O-STBC with code length P for N_T transmit antennas, which consists of K sets of dispersion matrices denoted as $\{\tilde{\mathbf{A}}_i, \tilde{\mathbf{B}}_i\}$, $1 \leq i \leq K$. An MDC-QOSTBC with code length $2P$ for $2N_T$ transmit antennas, which consists of $2K$ sets of dispersion matrices denoted as $\{\mathbf{A}_i, \mathbf{B}_i\}$, $1 \leq i \leq 2K$, can be constructed using the following mapping rules:

$$\text{Rule 1: } \mathbf{A}_i = \begin{bmatrix} \tilde{\mathbf{A}}_i & \mathbf{0} \\ \mathbf{0} & \tilde{\mathbf{A}}_i \end{bmatrix}; \quad \text{Rule 3: } \mathbf{A}_{K+i} = \begin{bmatrix} j\tilde{\mathbf{B}}_i & \mathbf{0} \\ \mathbf{0} & j\tilde{\mathbf{B}}_i \end{bmatrix};$$

$$1 \leq i \leq K \quad (4.21)$$

$$\text{Rule 2: } \mathbf{B}_i = \begin{bmatrix} \mathbf{0} & j\tilde{\mathbf{A}}_i \\ j\tilde{\mathbf{A}}_i & \mathbf{0} \end{bmatrix}; \quad \text{Rule 4: } \mathbf{B}_{K+i} = \begin{bmatrix} \mathbf{0} & \tilde{\mathbf{B}}_i \\ \tilde{\mathbf{B}}_i & \mathbf{0} \end{bmatrix}.$$

Note that $\{\mathbf{A}_i, \mathbf{B}_i\}$ can be square or rectangular designs, as can $\{\tilde{\mathbf{A}}_i, \tilde{\mathbf{B}}_i\}$.

Proof of Construction 4.1: Based on the structure of the dispersion matrices $\{\tilde{\mathbf{A}}_i, \tilde{\mathbf{B}}_i\}$ of O-STBC specified in (2.2), it can be proved that the mapping rules in (4.21) result in a new set of dispersion matrices $\{\mathbf{A}_i, \mathbf{B}_i\}$ that satisfy the MDC-QOC in (4.1). For example, (4.1) (i) can be verified as follows:

$$\mathbf{A}_i^H \mathbf{A}_k + \mathbf{A}_k^H \mathbf{A}_i \qquad 1 \leq i \leq K, K < k \leq 2K$$

$$= \begin{bmatrix} \tilde{\mathbf{A}}_i & \mathbf{0} \\ \mathbf{0} & \tilde{\mathbf{A}}_i \end{bmatrix}^H \begin{bmatrix} j\tilde{\mathbf{B}}_{k-K} & \mathbf{0} \\ \mathbf{0} & j\tilde{\mathbf{B}}_{k-K} \end{bmatrix} + \begin{bmatrix} j\tilde{\mathbf{B}}_{k-K} & \mathbf{0} \\ \mathbf{0} & j\tilde{\mathbf{B}}_{k-K} \end{bmatrix}^H \begin{bmatrix} \tilde{\mathbf{A}}_i & \mathbf{0} \\ \mathbf{0} & \tilde{\mathbf{A}}_i \end{bmatrix}$$

$$= \begin{bmatrix} j\tilde{\mathbf{A}}_i^H \tilde{\mathbf{B}}_{k-K} & \mathbf{0} \\ \mathbf{0} & j\tilde{\mathbf{A}}_i^H \tilde{\mathbf{B}}_{k-K} \end{bmatrix} + \begin{bmatrix} -j\tilde{\mathbf{B}}_{k-K}^H \tilde{\mathbf{A}}_i & \mathbf{0} \\ \mathbf{0} & -j\tilde{\mathbf{B}}_{k-K}^H \tilde{\mathbf{A}}_i \end{bmatrix}$$

$$= j\begin{bmatrix} \tilde{\mathbf{A}}_i^H \tilde{\mathbf{B}}_{k-K} - \tilde{\mathbf{B}}_{k-K}^H \tilde{\mathbf{A}}_i & \mathbf{0} \\ \mathbf{0} & \tilde{\mathbf{A}}_i^H \tilde{\mathbf{B}}_{k-K} - \tilde{\mathbf{B}}_{k-K}^H \tilde{\mathbf{A}}_i \end{bmatrix}$$

$$= \mathbf{0}.$$

where the property $\tilde{\mathbf{A}}_i^H \tilde{\mathbf{B}}_k - \tilde{\mathbf{B}}_k^H \tilde{\mathbf{A}}_i = \mathbf{0}$ for $i \neq k$ from (2.2) (iii) has been invoked.

The detailed verifications of the other conditions in (4.1) are omitted here as they are routine.   ∎

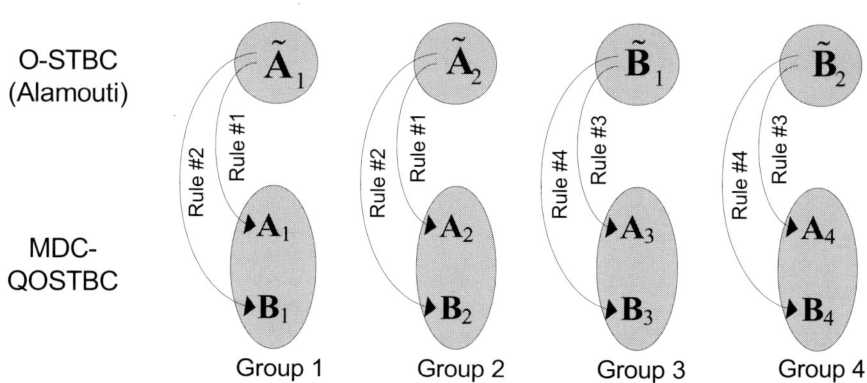

Fig. 4.3 Construction of MDC-QOSTBC from O-STBC

A graphical illustration for the construction of a four-antenna MDC-QOSTBC from the two-antenna Alamouti O-STBC is shown in Fig. 4.3, where $\tilde{\mathbf{A}}_1, \tilde{\mathbf{B}}_1, \tilde{\mathbf{A}}_2, \tilde{\mathbf{B}}_2$ denote the dispersion matrices (1.12) of the Alamouti O-STBC, and $\mathbf{A}_1$ to $\mathbf{A}_4$, $\mathbf{B}_1$ to $\mathbf{B}_4$ denote the dispersion matrices of the constructed MDC-QOSTBC. The codeword $\mathbf{C}$ of the resultant MDC-QOSTBC is:

$$\mathbf{C} = \begin{bmatrix} c_1^R + jc_3^R & c_2^R + jc_4^R & -c_1^I + jc_3^I & -c_2^I + jc_4^I \\ -c_2^R + jc_4^R & c_1^R - jc_3^R & c_2^I + jc_4^I & -c_1^I - jc_3^I \\ -c_1^I + jc_3^I & -c_2^I + jc_4^I & c_1^R + jc_3^R & c_2^R + jc_4^R \\ c_2^I + jc_4^I & -c_1^I - jc_3^I & -c_2^R + jc_4^R & c_1^R - jc_3^R \end{bmatrix}. \tag{4.22}$$

To compute the ML decoding metric of the above MDC-QOSTBC, let us reproduce the received signal model in (1.6) below:

$$r_p^{(k)} = \sqrt{\frac{E_s}{N_T}} \sum_{i=1}^{N_T} \left(h_{i,k} x_p^{(i)} + \eta_p^{(k)} \right). \tag{4.23}$$

Assuming that perfect channel state information (CSI) is available, the receiver computes the ML decision metric as follows:

$$\sum_{k=1}^{N_R}\sum_{p=1}^{P}\left|r_p^{(k)}-\sum_{i=1}^{N_T}h_{i,k}x_p^{(i)}\right|^2,\qquad(4.24)$$

where $x_p^{(i)}$, the transmitted symbol at i^{th} transmit antenna at p^{th} symbol period, corresponds to the pth row and i^{th} column entry of the MDC-QOSTBC codeword $\mathbf{C}$ in (4.22). The ML decision metric (4.24) hence becomes:

$$\sum_{k=1}^{N_R}\left\{\begin{aligned}&\left|r_1^{(k)}-\left[\begin{aligned}&h_{1,1}\left(c_1^{\text{R}}+jc_3^{\text{R}}\right)+h_{2,1}\left(c_2^{\text{R}}+jc_4^{\text{R}}\right)+\\&h_{3,1}\left(-c_1^{\text{I}}+jc_3^{\text{I}}\right)+h_{4,1}\left(-c_2^{\text{I}}+jc_4^{\text{I}}\right)\end{aligned}\right]\right|^2+\\[4pt]&\left|r_2^{(k)}-\left[\begin{aligned}&h_{1,2}\left(-c_2^{\text{R}}+jc_4^{\text{R}}\right)+h_{2,2}\left(c_1^{\text{R}}-jc_3^{\text{R}}\right)+\\&h_{3,2}\left(c_2^{\text{I}}+jc_4^{\text{I}}\right)+h_{4,2}\left(-c_1^{\text{I}}-jc_3^{\text{I}}\right)\end{aligned}\right]\right|^2+\\[4pt]&\left|r_3^{(k)}-\left[\begin{aligned}&h_{1,3}\left(-c_1^{\text{I}}+jc_3^{\text{I}}\right)+h_{2,3}\left(-c_2^{\text{I}}+jc_4^{\text{I}}\right)+\\&h_{3,3}\left(c_1^{\text{R}}+jc_3^{\text{R}}\right)+h_{4,3}\left(c_2^{\text{R}}+jc_4^{\text{R}}\right)\end{aligned}\right]\right|^2+\\[4pt]&\left|r_4^{(k)}-\left[\begin{aligned}&h_{1,4}\left(c_2^{\text{I}}+jc_4^{\text{I}}\right)+h_{2,4}\left(-c_1^{\text{I}}-jc_3^{\text{I}}\right)+\\&h_{3,4}\left(-c_2^{\text{R}}+jc_4^{\text{R}}\right)+h_{4,4}\left(c_1^{\text{R}}-jc_3^{\text{R}}\right)\end{aligned}\right]\right|^2\end{aligned}\right\},\qquad(4.25)$$

which can eventually be calculated as the sum $f_1+f_2+f_3+f_4$, where the functions f_1 to f_4 are given in (4.26). Since each f_i is just a function of a single complex symbol c_i, the minimization of the ML metric (4.25) is equivalent to minimizing the four f_i terms separately. This leads to a much lower decoding complexity than many existing QO-STBC whose ML metrics typically rely on two complex symbols, as shown earlier in (2.23).

$$f_1(c_1) = \sum_{k=1}^{N_R}\left[\left(\sum_{i=1}^{4}|h_{i,k}|^2\right)\left(|c_1^R|^2 + |c_1^I|^2\right) + 2\,\mathrm{Re}\left\{c_1^R(\alpha) - c_1^I(\beta) - c_1^R c_1^I(\gamma)\right\}\right];$$

$$f_2(c_2) = \sum_{k=1}^{N_R}\left[\left(\sum_{i=1}^{4}|h_{i,k}|^2\right)\left(|c_2^R|^2 + |c_2^I|^2\right) + 2\,\mathrm{Re}\left\{c_2^R(\chi) - c_2^I(\delta) - c_2^R c_2^I(\varphi)\right\}\right];$$

$$f_3(c_3) = \sum_{k=1}^{N_R}\left[\left(\sum_{i=1}^{4}|h_{i,k}|^2\right)\left(|c_3^R|^2 + |c_3^I|^2\right) + 2\,\mathrm{Re}\left\{jc_3^R(\alpha) + jc_3^I(\beta) + c_3^R c_3^I(\gamma)\right\}\right];$$

$$f_4(c_4) = \sum_{k=1}^{N_R}\left[\left(\sum_{i=1}^{4}|h_{i,k}|^2\right)\left(|c_4^R|^2 + |c_4^I|^2\right) + 2\,\mathrm{Re}\left\{jc_4^R(\chi) + jc_4^I(\delta) + c_4^R c_4^I(\varphi)\right\}\right],$$

(4.26)

where

$$\alpha = -h_{1,k}r_1^{(k)*} - h_{2,k}^* r_2^{(k)} - h_{3,k}r_3^{(k)*} - h_{4,k}^* r_4^{(k)},$$

$$\beta = -h_{3,k}r_1^{(k)*} - h_{4,k}^* r_2^{(k)} - h_{1,k}r_3^{(k)*} - h_{2,k}^* r_4^{(k)},$$

$$\gamma = 2\,\mathrm{Re}(h_{1,k}h_{3,k}^* + h_{2,k}h_{4,k}^*),$$

$$\chi = -h_{2,k}r_1^{(k)*} + h_{1,k}^* r_2^{(k)} - h_{4,k}r_3^{(k)*} + h_{3,k}^* r_4^{(k)},$$

$$\delta = -h_{4,k}r_1^{(k)*} + h_{3,k}^* r_2^{(k)} - h_{2,k}r_3^{(k)*} + h_{1,k}^* r_4^{(k)},$$

$$\varphi = 2\,\mathrm{Re}(h_{1,k}h_{3,k}^* + h_{2,k}h_{4,k}^*),$$

$h_{i,k}$ = spatial channel coefficient between the ith transmit and the kth receive antennas,

$r_p^{(k)}$ = received signals at the kth receive antenna at time instant p.

4.3.2 *Performance optimization*

Similar to the QO-STBCs, MDC-QOSTBC constructed by *Construction 4.1* or from a Preferred AOD Pair cannot achieve full transmit diversity directly. We therefore use the CR (constellation rotation) technique to help it attain full diversity and optimize its decoding performance. Note that GCLT (group-constrained linear transformation) and CR are equivalent for MDC-QOSTBC. This is because MDC-QOSTBC has only two real symbols per group, so its GCLT optimization is a two-dimensional problem which reduces to CR.

4.3.2.1 *Diversity product of MDC-QOSTBC*

We now derive the diversity product expression (defined in (1.23)) of MDC-QOSTBC, and use it to search for the optimum CR angle. From the construction rules of MDC-QOSTBC in (4.21) and the properties of O-STBC dispersion matrices in (2.2), we can obtain the following expression for the codeword distance matrix $\mathbf{A}_{CE}$ (defined in (1.21)) of MDC-QOSTBC:

$$
\begin{aligned}
\mathbf{A}_{CE} &= \left(\mathbf{C}-\mathbf{E}\right)^{H}\left(\mathbf{C}-\mathbf{E}\right) \\[2mm]
&= \left[\sum_{i=1}^{K}\left(\Delta_i^{R}\mathbf{A}_i + j\Delta_i^{I}\mathbf{B}_i\right)\right]^{H}\left[\sum_{i=1}^{K}\left(\Delta_i^{R}\mathbf{A}_i + j\Delta_i^{I}\mathbf{B}_i\right)\right] \\[2mm]
&= \sum_{i=1}^{K}\left[\left(\Delta_i^{R}\right)^{2}\mathbf{A}_i^{H}\mathbf{A}_i\right] + \sum_{i=1}^{K}\sum_{k=1,k\neq i}^{K}\left[\Delta_i^{R}\Delta_k^{R}\left(\underbrace{\mathbf{A}_i^{H}\mathbf{A}_k + \mathbf{A}_k^{H}\mathbf{A}_i}_{=0\ \text{due to MDC-QOC}}\right)\right] + \\[2mm]
&\quad \sum_{i=1}^{K}\left[\left(\Delta_i^{I}\right)^{2}\mathbf{B}_i^{H}\mathbf{B}_i\right] + \sum_{i=1}^{K}\sum_{k=1,k\neq i}^{K}\left[\Delta_i^{I}\Delta_k^{I}\left(\underbrace{\mathbf{B}_i^{H}\mathbf{B}_k + \mathbf{B}_k^{H}\mathbf{B}_i}_{=0\ \text{due to MDC-QOC}}\right)\right] + \\[2mm]
&\quad \sum_{i=1}^{K}\left[j\Delta_i^{R}\Delta_i^{I}\left(\mathbf{A}_i^{H}\mathbf{B}_i - \mathbf{B}_i^{H}\mathbf{A}_i\right)\right] + \sum_{i=1}^{K}\sum_{k=1,k\neq i}^{K}\left[j\Delta_i^{R}\Delta_k^{I}\left(\underbrace{\mathbf{A}_i^{H}\mathbf{B}_k - \mathbf{B}_k^{H}\mathbf{A}_i}_{=0\ \text{due to MDC-QOC}}\right)\right] \\[2mm]
&= \sum_{i=1}^{K}\left[\left(\Delta_i^{R}\right)^{2}\mathbf{A}_i^{H}\mathbf{A}_i\right] + \sum_{i=1}^{K}\left[\left(\Delta_i^{I}\right)^{2}\mathbf{B}_i^{H}\mathbf{B}_i\right] + \sum_{i=1}^{K}\left[j\Delta_i^{R}\Delta_i^{I}\left(\mathbf{A}_i^{H}\mathbf{B}_i - \mathbf{B}_i^{H}\mathbf{A}_i\right)\right]
\end{aligned}
$$

$$= \sum_{i=1}^{K}\left[\left(\Delta_i^R\right)^2\begin{bmatrix} \mathbf{I}_{N_T/2} & \mathbf{0} \\ \mathbf{0} & \mathbf{I}_{N_T/2} \end{bmatrix}\right] + \sum_{i=1}^{K}\left[\left(\Delta_i^I\right)^2\begin{bmatrix} \mathbf{I}_{N_T/2} & \mathbf{0} \\ \mathbf{0} & \mathbf{I}_{N_T/2} \end{bmatrix}\right] +$$

$$\sum_{i=1}^{K/2}\left[j\Delta_i^R\Delta_i^I\begin{bmatrix} \mathbf{0} & j2\mathbf{I}_{N_T/2} \\ j2\mathbf{I}_{N_T/2} & \mathbf{0} \end{bmatrix}\right] +$$

$$\sum_{i=K/2+1}^{K}\left[j\Delta_i^R\Delta_i^I\begin{bmatrix} \mathbf{0} & -j2\mathbf{I}_{N_T/2} \\ -j2\mathbf{I}_{N_T/2} & \mathbf{0} \end{bmatrix}\right]$$

$$= \alpha\begin{bmatrix} \mathbf{I}_{N_T/2} & \mathbf{0} \\ \mathbf{0} & \mathbf{I}_{N_T/2} \end{bmatrix} + \beta\begin{bmatrix} \mathbf{0} & \mathbf{I}_{N_T/2} \\ \mathbf{I}_{N_T/2} & \mathbf{0} \end{bmatrix},$$

(4.27)

where $\mathbf{C}$ and $\mathbf{E}$ represent two different codewords, Δ_i represents the possible error of symbols c_i, and

$$\alpha = \sum_{i=1}^{K}\left(\Delta_i^R\right)^2 + \left(\Delta_i^I\right)^2,$$

$$\beta = 2\sum_{i=1}^{K/2}\left(-\Delta_i^R\Delta_i^I + \Delta_{K/2+i}^R\Delta_{K/2+i}^I\right).$$

(4.28)

The determinant of the codeword distance matrix in (4.27) can be shown to be:

$$\det = \left[(\alpha+\beta)(\alpha-\beta)\right]^{\frac{N_T}{2}},$$

(4.29)

and its diversity product , ζ, is:

$$\zeta = \frac{1}{2\sqrt{N_T}}\underset{\mathbf{C}\neq\mathbf{E}}{\mathrm{Min}}\left[\det\right]^{\frac{1}{2P}}$$

$$= \frac{1}{2\sqrt{N_T}}\underset{\mathbf{C}\neq\mathbf{E}}{\mathrm{Min}}\left[\left[(\alpha+\beta)(\alpha-\beta)\right]^{\frac{N_T}{2}}\right]^{\frac{1}{2P}}.$$

(4.30)

In order to achieve full diversity and optimum coding gain, the diversity product has to be non-zero and its value has to be maximized. Following [16,31,32,33], when considering the minimum value of the determinant in (4.29), we can assume that only one of the symbols (e.g. symbol c_i) makes an error and the rest of the symbols are error free, hence the minimum (worst-case) determinant value in (4.30) can be simplified to:

$$\det = \left[\left((\Delta_i^R)^2 + (\Delta_i^I)^2 + 2\Delta_i^R \Delta_i^I\right)\left((\Delta_i^R)^2 + (\Delta_i^I)^2 - 2\Delta_i^R \Delta_i^I\right)\right]^{\frac{N_T}{2}}$$
$$= \left[\left(\Delta_i^R\right)^2 - \left(\Delta_i^I\right)^2\right]^{N_T}. \tag{4.31}$$

4.3.2.2 Optimum CR angle for square- and rectangular-QAM

The determinant (4.31) and the diversity product (4.30) may be zero due to possible occurrence of $|\Delta^R| = |\Delta^I|$ in most standard constellations such as QPSK and 16QAM. Hence CR technique must be used to avoid this situation. Assuming that c_i is now the rotated symbols, then the errors in c_i, Δ_i^R and Δ_i^I, can be expressed in terms of the error of the un-rotated constellation symbol, Λ_i, using the following relationship:

$$\Delta_i^R = \Lambda_i^R \cos\theta - \Lambda_i^I \sin\theta,$$
$$\Delta_i^I = \Lambda_i^R \sin\theta + \Lambda_i^I \cos\theta, \tag{4.32}$$

where θ represents the CR angle. Hence the determinant expression in (4.31) becomes (4.33):

$$\det = \left[\left(\Lambda_i^R\right)^2 \cos(2\theta) - 2\Lambda_i^R \Lambda_i^I \sin(2\theta) - \left(\Lambda_i^I\right)^2 \cos(2\theta)\right]^{N_T}. \tag{4.33}$$

Note that (4.33) is common to all the symbols c_i, hence the optimum θ obtained for it applies to all c_i symbols.

<u>*Theorem 4.4*</u>: The optimum CR angle of the MDC-QOSTBC with square- and rectangular-QAM is

$$\theta_{\text{opt}} = \frac{1}{2} \tan^{-1}(\frac{1}{2}) \tag{4.34}$$

for all code symbols, and the corresponding diversity product is:

$$\zeta = \frac{1}{2\sqrt{N_T}} \left(\frac{4}{5}\right)^{\frac{N_T}{4P}} d_{min}^{N_t/P},\tag{4.35}$$

where d_{min} is the minimum Euclidean distance between two QAM constellation points.

Proof of Theorem 4.4: For square- and rectangular-QAM constellation, the determinant expression of MDC-QOSTBC in (4.33) is similar to that of QO-STBC with GCLT in (3.32), hence the optimum angle of rotation is the same as in (3.37), i.e. $\theta_{opt} = \frac{1}{2} \tan^{-1}(\frac{1}{2}) = 13.28^0$. Note that this θ_{opt} value is the same for all code symbols in the MDC-QOSTBC codeword. This may be surprising as CR angle usually refers to the rotation of one complex symbol with respect to another complex symbol. For MDC-QOSTBC, it turns out that it is the relative angle between the I and Q components of a code symbol (treated as two real symbols) that matters. After all, MDC-QOSTBC is single–symbol–decodable. By substituting θ_{opt} and $\Lambda_i^R = \Lambda_i^I = d_{min}$ into (4.33) we get:

$$\det{}_{min} = \left(\frac{4}{5}\right)^{\frac{N_T}{2}} d_{min}^{2N_t},$$

$$\zeta = \frac{1}{2\sqrt{N_T}} \left(\frac{4}{5}\right)^{\frac{N_T}{4P}} d_{min}^{N_T/P}.\tag{4.36}$$

Hence *Theorem 4.4* is proved. ∎

4.3.2.3 *Optimum CR angle for PSK*

Unlike QAM, the optimum CR angle for MDC-QOSTBC with PSK constellation is unfortunately not tractable, hence we rely on computer search to find it. As shown in Fig. 4.4, the optimum CR angle is 31.7^0 for QPSK, and 4.9^0 for 8PSK. Likewise, these values are common for all PSK symbols in the codeword.

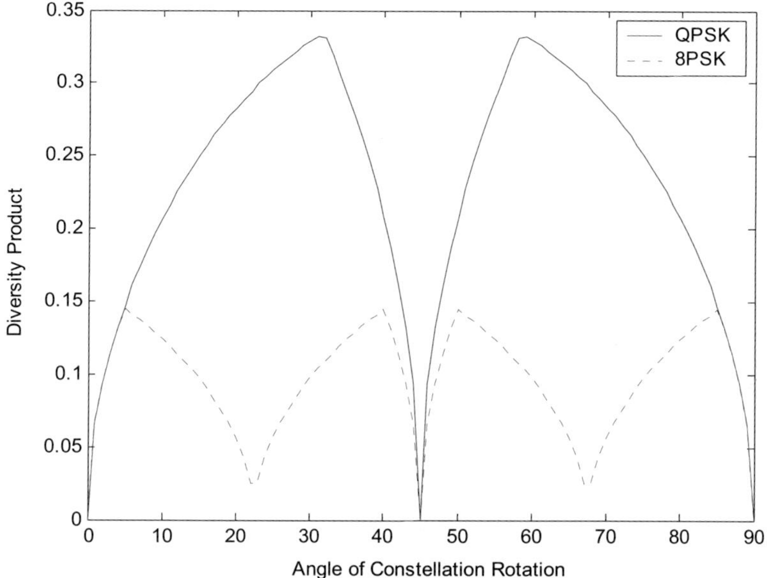

Fig. 4.4 Optimization of PSK rotation angle for MDC-QOSTBC

Table 4.5 compares the diversity product and ML decoding complexity (in terms of the number of real symbols required for joint detection (JD)) between O-STBC [17], QO-STBC with CR [16], QO-STBC with GCLT (Section 3.4), MDC-QOSTBC with optimize CR angle and CIOD [58] for 4-QAM and 8-PSK constellations. We can see that MDC-QOSTBC has a lower decoding complexity than the existing QO-STBC with CR. This decoding complexity reduction is traded-off by a slight reduction in the diversity product, which interestingly is also observed for CIOD and QO-STBC with GCLT. This appears to be a fundamental price to pay to obtain a lower decoding complexity. Although QO-STBC with GCLT achieves the same decoding complexity as MDC-QOSTBC, the latter supports *any* complex constellation, while GCLT is only applicable for square- or rectangular-QAM constellations. It is also noteworthy that for PSK constellations, MDC-QOSTBC has the same ML decoding complexity as O-STBC (since both require two real symbols for joint detection), but MDC-QOSTBC achieves a higher spectral efficiency of 3 bps/Hz because it has full code rate, while O-

STBC has a lower spectral efficiency of 2.25 bps/Hz due to a lower code rate of 3/4.

Table 4.5 Comparison of STBCs for four transmit antennas

STBC	4-QAM			8-PSK		
	Bps/Hz	Diver. Pdt.	No. of *real* syms for JD	Bps/Hz	Diver. Pdt.	No. of *real* syms for JD
Rate-3/4 O-STBC	1.5	0.4082	1	2.25	0.2209	2
QO-STBC CR [16]	2	0.3536	4	3	0.1674	4
MDC-QOSTBC (4.22)	2	0.3344	2	3	0.1453	2
QO-STBC GCLT	2	0.3344	2	N.A. for 8-PSK		
CIOD [58]	2	0.3344	2	Not reported		

4.3.3   *Non-square MDC-QOSTBC design*

In the previous section, we have only focused on the design of square MDC-QOSTBC. In this section, based on the construction method given in *Construction 4.1* (also including those square MDC-QOSTBC contructed by Preferred AOD Pair in section 4.2), we demonstrate how to make use of square or non-square O-STBC to construct non-square MDC-QOSTBC, which can give a higher code rate over a square design at the expense of a higher decoding delay (due to longer code length).

4.3.3.1   *MDC-QOSTBC for odd number of transmit antennas*

Construction 4.1 specifies how to construct MDC-QOSTBC for an even number of transmit antennas. In some cases, an odd number of antennas may be necessary. In this section, we will introduce a method to obtain MDC-QOSTBC for odd number of transmit antennas from a larger MDC-QOSTBC.

Lemma 4.1 [58, Theorem 4.3.15]: Let $\mathbf{M}$ be a Hermitian matrix with size m, and d be an integer with $1 \leq d \leq m$, and let $\mathbf{M}_d$ denote any d-by-d principal submatrix of $\mathbf{M}$ (obtained by deleting $m\text{-}d$ rows and the corresponding columns from $\mathbf{M}$). For each integer k in the range of $1 \leq k \leq d$, we have

$$\lambda_k(\mathbf{M}) \leq \lambda_k(\mathbf{M}_d) \leq \lambda_{k+m-d}(\mathbf{M}) \tag{4.37}$$

where $\lambda_k(\mathbf{M})$ represents the kth eigenvalue of matrix $\mathbf{M}$, ordered in increasing eigenvalues magnitude.

Theorem 4.5: By removing any column in the codeword of a full-diversity MDC-QOSTBC, the resultant codeword is still an MDC-QOSTBC that achieves full diversity and the same code rate.

Proof of Theorem 4.5: The removal of any column from a STBC can be represented as follows:

$$\overline{\mathbf{C}} = \mathbf{CT} \tag{4.38}$$

where $\overline{\mathbf{C}}$ of size $P \times \overline{N}_T$ is the new STBC after column removal, and $\mathbf{T}$ is a truncated version of an identity matrix after column removal at the same column positions as $\overline{\mathbf{C}}$ (hence $\mathbf{T}$ is an rectangular matrix of size $N_T \times \overline{N}_T$). For example, to remove the last column of $\mathbf{C}$ so as to reduce the number of transmit antennas from $N_T = 4$ to $\overline{N}_T = 3$, the following $\mathbf{T}$ may be used:

$$\mathbf{T} = \begin{bmatrix} 1 & 0 & 0 \\ 0 & 1 & 0 \\ 0 & 0 & 1 \\ 0 & 0 & 0 \end{bmatrix}. \tag{4.39}$$

In this case, the dispersion matrices of the new STBC $\overline{\mathbf{C}}$ would be:

$$\begin{aligned} \overline{\mathbf{A}}_i &= \mathbf{A}_i\mathbf{T}, \\ \overline{\mathbf{B}}_i &= \mathbf{B}_i\mathbf{T}. \end{aligned} \tag{4.40}$$

It can be easily verified that as long as the dispersion matrices $\{\mathbf{A}_i, \mathbf{B}_i\}$ satisfy the MDC-QOC in (4.1), $\{\overline{\mathbf{A}}_i, \overline{\mathbf{B}}_i\}$ will satisfy the MDC-QOC too. Furthermore, the new STBC will have the following codeword distance matrix:

$$\overline{\mathbf{A}}_{\mathbf{CE}} = \mathbf{T}^{\mathrm{H}}(\mathbf{C}-\mathbf{E})^{\mathrm{H}}(\mathbf{C}-\mathbf{E})\mathbf{T}$$

$$= \mathbf{T}^{\mathrm{H}}\mathbf{A}_{\mathbf{CE}}\mathbf{T} \tag{4.41}$$

$$= \overline{N}_T\text{-by-}\overline{N}_T \text{ principle submatrix of } \mathbf{A}_{\mathbf{CE}}.$$

Since $\overline{\mathbf{A}}_{\mathbf{CE}}$ and $\mathbf{A}_{\mathbf{CE}}$ are both Hermitian matrixes, *Lemma 4.1* applies. As a result, if $\mathbf{A}_{\mathbf{ce}}$ does not have any zero eigenvalue (since $\mathbf{C}$ is assumed to achieve full diversity), $\overline{\mathbf{A}}_{\mathbf{CE}}$ will not too (as all k^{th} eigenvalue of $\overline{\mathbf{A}}_{\mathbf{CE}}$ will be at least equal to the kth eigenvalue of $\mathbf{A}_{\mathbf{CE}}$), hence $\overline{\mathbf{C}}$ can also achieve full diversity. Therefore, *Theorem 4.5* is proved.   ∎

4.3.3.2 *Maximum code rate of square MDC-QOSTBC*

Construction 4.1 implies that the maximum achievable code rate of an MDC-QOSTBC is related to the maximum achievable code rate of an O-STBC. Specifically, an MDC-QOSTBC for $2N_T$ transmit antennas has $2K$ dispersion matrices, each of duration $2P$, hence its code rate is $2K / 2P = K/P$, which is the same as the code rate of the lower-order O-STBC used to generate it. Based on the known maximum achievable code rate property of O-STBC, the maximum achievable code rate of an MDC-QOSTBC constructed from *Construction 4.1* can be obtained as follows:

<u>*Theorem 4.6*</u>: The maximum achievable code rate of a square MDC-QOSTBC constructed from *Construction 4.1* is:

$$R_{\text{MDC-QOSTBC}} = \frac{2\log_2 k}{k}, \quad \text{where } k = 2^{\lceil \log_2 N_T \rceil}. \tag{4.42}$$

<u>*Proof of Theorem 4.6*</u>: In [17,19], it has been shown that a square O-STBC with a code length of $\tilde{N}_T$ has a maximum achievable code rate of

$$R_{\text{O-STBC}} = \frac{\log_2 i + 1}{i}, \quad \text{where } i = 2^{\lceil \log_2 \tilde{N}_T \rceil}. \tag{4.43}$$

Since an MDC-QOSTBC constructed from (4.21) has the same code rate as the half-size O-STBC used to construct it, we get the maximum achievable code rate of square MDC-QOSTBC by substituting $N_T = 2\tilde{N}_T$ into (4.43):

$$
\begin{aligned}
R_{\text{MDC-QOSTBC}} &= \frac{\log_2 2^{\lceil \log_2(N_T/2) \rceil} + 1}{2^{\lceil \log_2(N_T/2) \rceil}} \\
&= \frac{2\log_2 2^{\lceil \log_2(N_T) \rceil}}{2^{\lceil \log_2(N_T) \rceil}} \\
&= \frac{2\log_2 k}{k}, \qquad \text{where } k = 2^{\lceil \log_2(N_T) \rceil}.
\end{aligned}
\tag{4.44}
$$

Hence *Theorem 4.6* is proved. ∎

Table 4.6 Comparison between square O-STBC and square QO-STBC

Tx. Antennas	O-STBC			QO-STBC			
	Ref.	P	R	Reference	P	R	Decoding Complexity
2	[5]	2	1	N.A.			
4	[17] [19]	4	3/4	QO-STBC with CR [16]	4	1	4
				CIOD / ACIOD [58,60]	4	1	2
				MDC-QOSTBC	4	1	2
8	[17] [19]	8	1/2	QO-STBC with CR [16]	8	3/4	4
				CIOD / ACIOD [58,60]	8	3/4	2
				MDC-QOSTBC	8	3/4	2
16	[17] [19]	16	1/2	QO-STBC with CR [16]	16	1/2	4
				CIOD / ACIOD [58,60]	16	1/2	2
				MDC-QOSTBC	16	1/2	2

In Table 4.6, we give a comparison of the code length P, maximum achievable code rate R and decoding complexity (i.e. the number of real symbols required for joint ML detection) of square full-diversity MDC-QOSTBC constructed using *Construction 4.1* versus the square O-STBC, full-diversity QO-STBC and full-diversity CIOD/ACIOD for various number of transmit antennas. The comparison shows that MDC-QOSTBC achieves:

– a higher code rate than O-STBC with the same diversity level (number of transmit antennas), just as QO-STBC do;
– a lower decoding complexity than many existing full-diversity QO-STBC with CR at the same code rate;
– the same code rate and decoding complexity as a full-diversity CIOD/ACIOD with CR (differences between them will be discussed in a later section).

From Table 4.6, we also notice that the maximum achievable code rates of square MDC-QOSTBC for four and eight transmit antennas are the same as the ones derived in *Theorem 4.2* based on the Preferred AOD Pair concept.

4.3.3.3 *Maximum code rate of non-square MDC-QOSTBC*

It has been shown in [28] that for more than 4 transmit antennas, a non-square O-STBC has a higher achievable code rate than a square O-STBC. We will next derive the maximum achievable code rate of a non-square MDC-QOSTBC constructed from non-square O-STBC using *Construction 4.1*.

<u>*Theorem 4.7*</u>: The maximum achievable code rate of a non-square MDC-QOSTBC constructed using *Construction 4.1* is:

$$R_{\text{MDC-QOSTBC}} = \frac{k+1}{2k}, \quad \text{where } k = \left\lceil \frac{N_T}{4} \right\rceil. \tag{4.45}$$

<u>*Proof of Theorem 4.7*</u>: It has been shown in [28] that a non-square O-STBC for $\tilde{N}_T$ antennas has the maximum achievable code rate of:

$$R_{\text{O-STBC}} = \frac{i+1}{2i}, \quad \text{where } i = \left\lceil \frac{\tilde{N}_T}{2} \right\rceil. \tag{4.46}$$

Following the proof for *Theorem 4.6*, *Theorem 4.7* can be easily proved by substituting $N_T = 2\tilde{N}_T$ into (4.46). $\blacksquare$

Table 4.7 Comparison between non-square O-STBC and non-square MDC-QOSTBC

Tx. Antennas	O-STBC [28]		MDC-QOSTBC from *Construction 4.1*		GCIOD [61]	
	P	R	P	R	P	R
2	2	1	N.A.		N.A.	
3-4	4	3/4	4	1	4	1
5	10	4/6	8	3/4	14	6/7
6	30	4/6				
7	56	5/8			35	4/5
8	56	5/8			50	4/5
9	210	6/10	20	4/6	No information is available	
10	420	6/10				
11-12	792	7/12	60	4/6		
13	3003	8/14	112	5/8		
14	6006	8/14				
15-16	11440	9/16				

A comparison between non-square O-STBC and MDC-QOSTBC is shown in Table 4.7. It shows that although the non-square MDC-QOSTBC has higher code rate, it has a higher code length too, which leads to longer decoding latency. For example, for 16 transmit antennas, the *non-square* MDC-QOSTBC (constructed from non-square O-STBC using *Construction 4.1*) has a code rate of 5/8 and a code length of 112, whiles the *square* MDC-QOSTBC (constructed from square O-STBC) has a slightly lower code rate of 4/8 but a much shorter code length of just 16. We also include in Table 4.7 comparisons with the Generalised CIOD (GCIOD) [61] which has the same decoding complexity as MDC-QOSTBC. It is observed that GCIOD achieves a higher code rate than MDC-QOSTBC by increasing the decoding delay too.

4.4 Performance Results

The optimum CR angle, minimum determinant (coding gain) and P_o (probability that an antenna transmits the "zero" symbol) values of QO-STBC, CIOD and MDC-QOSTBC with 4QAM constellation for four transmit antennas are compared in Table 4.8, while their bit error rate (BER) are compared in Fig. 4.5.

Table 4.8 Comparison of QO-STBCs for four transmit antennas

	Optimum Constellation Angle	No. of Real Symbols for ML Joint Detection	Minimum Determinant	P_o
QO-STBC with CR	45^0	4	0.3536	0
CIOD	31.72^0	2	0.3344	50%
MDC-QOSTBC (4.22)	13.29^0	2	0.3344	0

These results show that MDC-QOSTBC suffers a slightly reduced diversity product and a slight 0.5 dB loss at BLER of 10^{-4} compared to the existing QO-STBCs, which have higher decoding complexity. Interestingly, the same performance loss is also observed for CIOD that has the same decoding complexity as MDC-QOSTBC. And the performance gap remains around 0.5 dB between MDC-QOSTBC and QO-STBC for eight transmit antennas. Hence it appears to be a fundamental price to pay for lower decoding complexity. Next, comparing MDC-QOSTBC against CIOD, we observe that although they have almost identical decoding performance, the MDC-QOSTBC in (4.22) does not transmit any zero symbol (hence it achieves the ideal value of $P_o = 0$), while the CIOD transmits zero symbols on half of the antennas at any one time (hence $P_o = 50\%$). As transmitting zero symbols on an antenna means turning off that antenna and inducing high peak-to-average power ratio on the radio front-end, and turning off transmit antennas *regularly* generates undesirable low-frequence interference, having low P_o value is advantageous in terms of practical considerations.

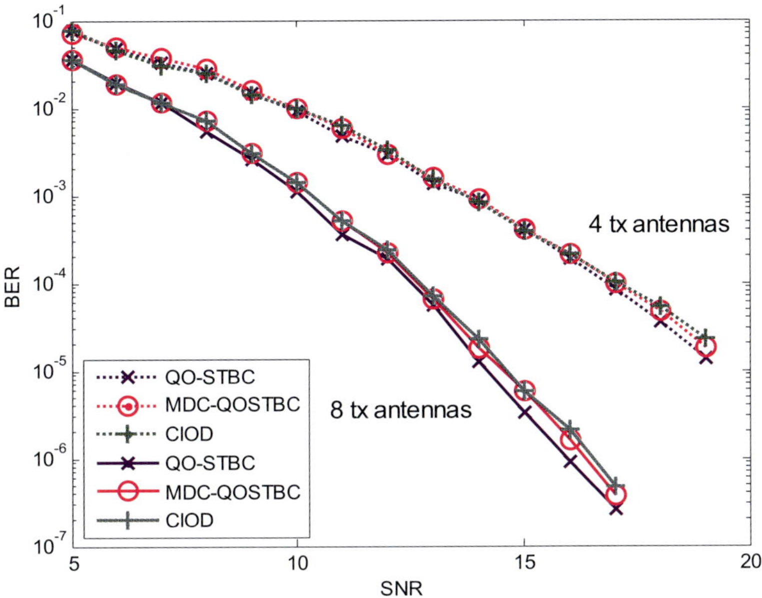

Fig. 4.5 BER performance of STBCs with spectral efficiency of 2 bps/Hz

For the cases of three and five transmit antennas, corresponding comparisons between MDC-QOSTBC, CIOD and ACIOD with 4QAM comparisons are presented in Table 4.9 and Fig. 4.6. the CIOD and MDC-QOSTBC for three transmit antennas are obtained by removing the last column of the codeword of their four-antenna counterparts, while the CIOD and MDC-QOSTBC for five transmit antennas are obtained by removing the first and last two columns from of the transmission matrix of their eight-antenna counterparts, based on the guideline given in [58]. The results in Table 4.9 and Fig. 4.6 show that MDC-QOSTBC can achieve higher diversity product and hence lower BLER than CIOD. Furthermore, MDC-QOSTBC performs comparably with ACIOD in terms of diversity product, but is more favourable in terms of P_o. Combined with earlier observations from Table 4.8 and Fig. 4.6, we can summarize that MDC-QOSTBC is more versatile in supporting both odd and even numbers of transmit antennas, while CIOD supports only an even number of transmit antennas well and ACIOD supports only an odd number of transmit antennas well.

Table 4.9 Comparison of STBCs for three and five transmit antennas

	Optimum Constellation Angle	3 Tx Antennas		5 Tx Antennas	
		Diversity Product	P_o	Diversity Product	P_o
CIOD [58]	31.72^0	0.2504	50%	0.2489	50%
ACIOD [60]	31.72^0	0.3564	33%	0.2946	20%
MDC-QOSTBC	13.29^0	0.3641	0	0.2996	0

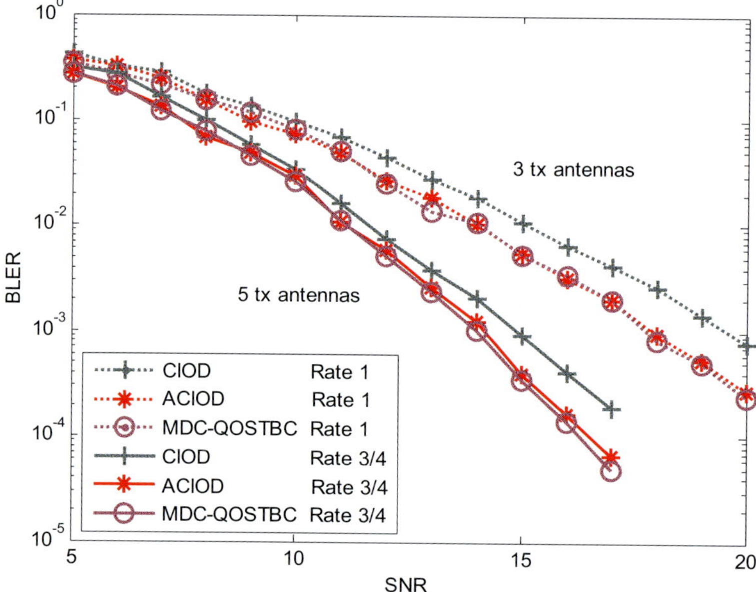

Fig. 4.6 Simulation performance of STBCs with 4-QAM constellation for three and five transmit antennas

4.5 Chapter Summary

In this chapter, we present the single–symbol–decodable MDC-QOSTBC as the QO-STBC with the smallest joint detection search space, and derive its unifying algebraic structure. By investigating the algebraic relationship between MDC-QOSTBC and AOD, we introduce the concept of the Preferred AOD Pair and show that a square MDC-QOSTBC can be constructed by two AODs that form a Preferred AOD Pair. The lower bound on the achievable code rate of MDC-QOSTBC constructed from a Preferred AOD Pair is derived to be 1 for four transmit antennas and 3/4 for eight transmit antennas. We present two systematic approaches to construct MDC-QOSTBC: one based on Quaternion matrices and Preferred AOD Pair, the other based on a set of dispersion matrix mapping rules involving O-STBC. The optimum CR angle for the MDC-QOSTBC to achieve full diversity and optimum coding gain is found to be 13.29^0 for square- or rectangular-QAM, 31.7^0 for QPSK, and 4.9^0 for 8PSK. In order to support odd number of transmit antennas, we show that columns of an MDC-QOSTBC codeword can be removed without affecting its diversity gain or decoding complexity. Compared with many existing QO-STBC with CR, the MDC-QOSTBC can achieve a lower ML decoding complexity without much penalty in coding gain or decoding performance. Compared with QO-STBC with GCLT, MDC-QOSTBC can support any complex constellation while QO-STBC with GCLT can only support square- and rectangular-QAM constellation. Compared with the CIOD and ACIOD, MDC-QOSTBC has better power distribution property and can better support different number of transmit antennas.

Chapter 5

Differential QO-STBC

The QO-STBC transmit diversity schemes discussed so far are designed for coherent detection, with the spatial channel gains and phases assumed available at the receiver via channel estimation. However, the complexity and cost of MIMO channel estimation grow very quickly with the number of transmit and receive antennas [62]. As a result, *Differential Space-Time Modulation* (DSTM) has been proposed to provide transmit diversity without requiring channel estimation. It is equivalent to the blind detection scheme mentioned in Section 1.1.

Similar to coherent transmit diversity, DSTM with high code rate and low decoding complexity is always desired. Existing DSTM schemes based on Orthogonal Space-Time Block Code (O-STBC) [63] have the lowest decoding complexity but cannot support full rate when more than two transmit antennas are used. On the other hand, other DSTM schemes that are not based on O-STBC, such as those reported in [62,64-70], have high decoding complexity involving large joint detection search space. Hence a DSTM scheme that can achieve both high code rate and low decoding complexity is again called for. We address these challenges by constructing DSTM schemes based on single- and double–symbol–decodable QO-STBC.

5.1 DSTM Codeword Model and Design Criteria

In order to be consistent with the signal models commonly adopted in the literature on this topic [63,71,72], in this chapter we employ a slightly different signal model from the one used in Chapter 2.

112

Most differential space-time modulations are designed based on a set of unitary or quasi-unitary matrices [62,63-65,71,72]. For the DSTM schemes described in [63,71,72], the set of unitary or quasi-unitary matrices $\{\mathbf{U}_i\}$ are constructed from a STBC, whose codeword $\mathbf{C}$ is represented as follows:

$$\mathbf{C} = \frac{1}{\sqrt{K}}\sum_{i=1}^{K}\left(c_i^{\mathrm{R}}\mathbf{A}_i + jc_i^{\mathrm{I}}\mathbf{B}_i\right).\qquad(5.1)$$

Note that (5.1) and (1.10) are similar except for the factor $1/\sqrt{K}$ in (5.1), which is used to normalize the STBC codeword matrices to be unitary or quasi-unitary. Also, the STBC codeword $\mathbf{C}$ in (5.1) and the dispersion matrices $\mathbf{A}$ and $\mathbf{B}$ are of size $N_T \times P$ rather than $P \times N_T$ as in (1.10), where N_T and P are the number of transmit antennas and code length, respectively.

The error probability performance of a unitary DSTM scheme with optimal differential decoder has been analyzed in [62]. Specifically, the diversity gain that can be achieved by DSTM is defined as:

$$\min_{k \neq l}\left[\mathrm{Rank}\left(\mathbf{U}_k - \mathbf{U}_l\right)\right],\qquad(5.2)$$

where $\mathbf{U}_k$ and $\mathbf{U}_l$ are two unitary codewords. To achieve full transmit diversity, the minimum rank in (5.2) must be equal to N_T.

For a full-rank unitary DSTM code, its *coding gain* is defined in [62,63] as

$$\min_{k \neq l}\left[N_T \times \mathrm{Det}\left(\left(\mathbf{U}_k - \mathbf{U}_l\right)\left(\mathbf{U}_k - \mathbf{U}_l\right)^{\mathrm{H}}\right)^{1/N_T}\right].\qquad(5.3)$$

To achieve minimum error probability, the coding gain in (5.3) must be maximized. It is clear that the diversity gain in (5.2) is the same as that defined in Section 1.5, while the coding gain in (5.3) is analogous to the diversity product defined in (1.23).

5.2 Unitary DSTM Based on QO-STBC

5.2.1 Literature review

Hochwald and Marzetta proposed the use of unitary space-time codes to construct a unitary DSTM scheme, in which the signals transmitted by different antennas are mutually orthogonal [66,67]. Optimal receivers, error bounds, and design criteria for unitary DSTM codes were derived in [66], while some specific code constructions were given in [67]. Hughes designed a DSTM based on group codes [62,68]. A similar idea was also reported in [65]. Jing and Hassibi [64] proposed a DSTM scheme based on Sp(2), which is able to make use of the simpler sphere decoding (SD) algorithm. Tarokh and Jafarkhani proposed a scheme based on the O-STBC [69,70], so did Ganesan and Stoica [63], though with a much simpler decoder. Among these DSTM schemes, the scheme in [63] is single-symbol decodable, hence it has the lowest decoding complexity. However, its maximum achievable code rate is limited to 3/4 for four transmit antennas and 1/2 for eight transmit antennas as it is based on O-STBC. On the other hand, all the other DSTM schemes have higher decoding complexity as their decoders need to search over a larger search space.

To strike a better balance between code rate and decoding complexity, we had proposed in [73] and [74,75] respectively to construct DSTM scheme using unitary matrices derived from double–symbol–decodable QO-STBC, or using quasi-unitary matrices derived from single–symbol–decodable MDC-QOSTBC. Their design criteria, decoding complexity and decoding performance will be elaborated in this chapter.

5.2.2 Signal model of unitary DSTM scheme

Consider a MIMO communication system with N_T transmit and N_R receive antennas. Let $\mathbf{H}_p$ be the $N_R \times N_T$ fading channel gain matrix at a time p. The $(i, k)^{\text{th}}$ element of $\mathbf{H}_p$ is the channel coefficient for the signal path from the kth transmit antenna to the ith receive antenna. Let $\mathbf{X}_p$ be the

$N_T \times P$ DSTM codeword transmitted at time p. Then, the received $N_R \times P$ signal matrix $\mathbf{R}_p$ can be written as

$$\mathbf{R}_p = \mathbf{H}_p \mathbf{X}_p + \mathbf{N}_p, \tag{5.4}$$

where $\mathbf{N}_p$ represents the additive white Gaussian channel noise. In this chapter, the code length P is set equal to N_T as in [62,63], so that the transmitted codeword $\mathbf{X}_p$ is a square matrix.

At the start of transmission, we transmit a *known* codeword $\mathbf{X}_0$, which is a unitary matrix of size $N_T \times N_T$. The purpose of $\mathbf{X}_0$ is to initalise the transmission and set as reference for subsequent transmission, it does not carry information. The info-carrying codeword $\mathbf{X}_p$ transmitted at a time p is differentially encoded by

$$\mathbf{X}_p = \mathbf{X}_{p-1} \mathbf{U}_p, \tag{5.5}$$

where $\mathbf{U}_p$ is a unitary matrix (such that $\mathbf{U}_p \mathbf{U}_p^{\mathrm{H}} = \mathbf{I}$), called the *code matrix*, that contains information of the transmitted data. Since $\mathbf{X}_0$ and $\mathbf{U}_p$ are both unitary, $\mathbf{X}_p$ will be unitary at all time. Hence the requirement for the code matrix $\mathbf{U}_p$ to be unitary ensures that all the transmitted codewords have constant power at all times. If we assume that the channel remains unchanged during two consecutive code periods, i.e. $\mathbf{H}_p = \mathbf{H}_{p-1}$, the received signal $\mathbf{R}_p$ at a time p can be expressed [63] as

$$\begin{aligned}
\mathbf{R}_p &= \mathbf{H}_p \mathbf{X}_{p-1} \mathbf{U}_p + \mathbf{N}_p \\
&= (\mathbf{R}_{p-1} - \mathbf{N}_{p-1}) \mathbf{U}_p + \mathbf{N}_p \\
&= \mathbf{R}_{p-1} \mathbf{U}_p + \tilde{\mathbf{N}}_p.
\end{aligned} \tag{5.6}$$

Based on (5.6), the received signal $\mathbf{R}_p$ can be differentially decoded as it depends only on the previous received signal block $\mathbf{R}_{p-1}$, the code matrix $\mathbf{U}_p$ and an equivalent additive white Gaussian noise $\tilde{\mathbf{N}}_p = -\mathbf{N}_{p-1} \mathbf{U}_p + \mathbf{N}_p$. Since $\mathbf{N}_p$ and $\mathbf{N}_{p-1}$ are white and $\mathbf{U}_p$ is unitary, it is shown in [64] that the equivalent noise $\tilde{\mathbf{N}}_p$ is white. Hence the decision metric for (5.6) is

$$\begin{aligned}
\hat{\mathbf{U}}_p &= \arg \min_{\mathbf{U}_p \in \mathcal{U}} \mathrm{Tr}\left(\left\{ \mathbf{R}_p - \mathbf{R}_{p-1} \mathbf{U}_p \right\}^{\mathrm{H}} \left\{ \mathbf{R}_p - \mathbf{R}_{p-1} \mathbf{U}_p \right\} \right) \\
&= \arg \max_{\mathbf{U}_p \in \mathcal{U}} \mathrm{Re}\left\{ \mathrm{Tr}\left(\mathbf{R}_p^{\mathrm{H}} \mathbf{R}_{p-1} \mathbf{U}_p \right) \right\},
\end{aligned} \tag{5.7}$$

where $\mathrm{Tr}(\mathbf{M})$ represents the trace of a matrix $\mathbf{M}$ and $\mathcal{U}$ denotes the set of all possible code matrices $\mathbf{U}_p$.

5.2.3 *Double–symbol–decodable unitary DSTM*

5.2.3.1   Unitary DSTM Based on Double–Symbol–Decodable QO-STBC

In this section, we present a new unitary DSTM scheme with low decoding complexity which is developed based on square double–symbol–decodable QO-STBC [73]. Let us use the QO-STBC J4 for four transmit antennas in (2.19) as an example. It has the following property:

$$(\mathbf{J4})(\mathbf{J4})^{\mathrm{H}} = \frac{1}{4}\begin{bmatrix} \alpha & 0 & 0 & \beta \\ 0 & \alpha & -\beta & 0 \\ 0 & -\beta & \alpha & 0 \\ \beta & 0 & 0 & \alpha \end{bmatrix},$$

(5.8)

where

$$\alpha = \sum_{i=1}^{4}|c_i|^2,$$

$$\beta = 2\,\mathrm{Re}\!\left(c_1 \times c_4^* - c_2 \times c_3^*\right),$$

(5.9)

and c_i are the code symbols in **J4**.

To use **J4** as the unitary code matrix of a DSTM scheme, its α and β values in (5.8) must be equal to 4 and 0 respectively. However it is clear from (5.9) that these cannot be achieved if c_1 to c_4 are independent PSK or QAM symbols.

To have $\beta = 0$, we can see from (5.9) that $\mathrm{Re}(c_1 c_4^*)$ must be equal to $\mathrm{Re}(c_2 c_3^*)$. This suggests that c_1 and c_4 should be jointly mapped to a symbol-pair $\{a_i, b_i\}$, while c_2 and c_3 should be mapped to another symbol-pair $\{a_k, b_k\}$, such that $\mathrm{Re}(a_i b_i^*) = \mathrm{Re}(a_k b_k^*)$ for all i and k. To further achieve $\alpha = 4$, we require in (5.9) that the symbol-pairs must further be such that $|a_i|^2 + |b_i|^2 + |a_k|^2 + |b_k|^2 = 4$, or $|a_i|^2 + |b_i|^2 = |a_k|^2 + |b_k|^2 = 2$ if all symbol-pairs are assumed to have equal power. The collection of all symbol-pairs $\{a_i, b_i\}$, denoted as the joint constellation set $\mathcal{M}$, must hence satisfy the constraints $\mathrm{Re}(a_i b_i^*) = v$ where v is any constant value, and $|a_i|^2 + |b_i|^2 = 2$. To maximize the coding gain of the DSTM scheme, $\mathcal{M}$ should further be designed to maximize the worst-case determinant value of the codeword distance matrix.

The determinant of the codeword distance matrix of **J4** has been shown in (2.29) to be:

$$\det = \frac{1}{4}\left[\left(|\Delta_1 + \Delta_4|^2 + |\Delta_2 - \Delta_3|^2\right) \times \left(|\Delta_1 - \Delta_4|^2 + |\Delta_2 + \Delta_3|^2\right)\right]^2, \quad (5.10)$$

where $\Delta_i = c_i - e_i$, $1 \le i \le 4$, represents the possible error in c_i when the receiver decides erroneously in favor of e_i assuming that c_i was transmitted. As explained in Section 2.2.2.1, when considering the minimum determinant value of (5.10) we can assume half of the codeword errors to be zero (e.g. $\Delta_2 = \Delta_3 = 0$), hence the minimum value of (5.10) can be simplified as:

$$\det = \frac{1}{4}\left[\left(|\Delta_1 + \Delta_4|^2\right) \times \left(|\Delta_1 - \Delta_4|^2\right)\right]^2 \quad (5.11)$$

assuming $\Delta_2 = \Delta_3 = 0$. In order to achieve full transmit diversity and maximum coding gain, the value in (5.11) has to be non-zero and maximized.

Summarizing the above, the design requirements on the joint constellation set $\mathcal{M}$, which consists of L sets of complex-valued symbol-pair $\{a_i, b_i\}$, for $1 \le i \le L$, are:

$$\text{(i) Unitary Criterion: } \mathrm{Re}(a_i b_i^*) = v;$$

$$\text{(ii) Power Criterion: } |a_i|^2 + |b_i|^2 = 2; \quad (5.12)$$

(iii) Performance Criterion:

$$\underset{\arg a_i, b_i}{\text{maximize}} \ \min\left\{\left[|\Delta a_{ik} + \Delta b_{ik}|^2 \times |\Delta a_{ik} - \Delta b_{ik}|^2\right]^2\right\},$$

where v can be any constant value, and $\Delta a_{ik} = a_i - a_k$, $\Delta b_{ik} = b_i - b_k$ for all $1 \le i \ne k \le L$. The spectral efficiency, R, of the resultant DSTM scheme based on full-rate QO-STBC is $R = 2(\log_2 L)/N_T$ bps/Hz.

As an example, consider a system with four transmit antennas ($N_T = 4$) and a desired spectral efficiency of $R = 2$ bps/Hz. From the above expression of R, the required constellation size is $L = 16 = 2^4$. Hence, in the *encoding process*, we will map 4 data bits to one of the 16 symbol-pairs $\{a_i, b_i\}$ in $\mathcal{M}$ to form the code symbols $\{c_1, c_4\}$ in **J4**, and then we map another 4 data bits to the symbol-pair $\{a_k, b_k\}$ in $\mathcal{M}$ to form the code symbols $\{c_2, c_3\}$ in **J4**. The resultant **J4** with c_1 to c_4 assigned as above will be unitary, and it can now be used as the DSTM code matrix $\mathbf{U}_p$ in (5.5).

In the *decoding process*, with $\mathbf{U}_p$ in (5.5) set to **J4**, the decision metrics in (5.7) can be simplified to:

$$
\begin{aligned}
\hat{\mathbf{U}}_p &= \arg \underset{\mathbf{U}_p \in \mathcal{U}}{\operatorname{Max}} \operatorname{Re}\left\{ \operatorname{Tr}\left(\mathbf{R}_p^{\mathrm{H}} \mathbf{R}_{p-1} \mathbf{U}_p \right) \right\} \\
&= \arg \underset{c_1,\ldots,c_4 \in \mathcal{M}}{\operatorname{Max}} \operatorname{Re}\left\{ \operatorname{Tr}\left[\mathbf{R}_p^{\mathrm{H}} \mathbf{R}_{p-1} \left(\sum_{i=1}^{4} \left(c_i^{\mathrm{R}} \mathbf{A}_i + j c_i^{\mathrm{I}} \mathbf{B}_i \right) \right) \right] \right\} \\
&= \arg \underset{c_1,\ldots,c_4 \in \mathcal{M}}{\operatorname{Max}} \sum_{i=1}^{4} \operatorname{Re}\left\{ \operatorname{Tr}\left[\mathbf{R}_p^{\mathrm{H}} \mathbf{R}_{p-1} \left(c_i^{\mathrm{R}} \mathbf{A}_i + j c_i^{\mathrm{I}} \mathbf{B}_i \right) \right] \right\},
\end{aligned} \tag{5.13}
$$

where $\mathbf{A}$ and $\mathbf{B}$ are the dispersion matrices of **J4**.

Since c_1 and c_4 are modulated independently from c_2 and c_3, (5.13) can be separated into two equations as follows:

$$
\begin{aligned}
\{\hat{c}_1, \hat{c}_4\} &= \arg \underset{\{c_1, c_4\} \in \mathcal{M}}{\operatorname{Max}} \left[\sum_{i=1,4} \left[\begin{array}{l} \operatorname{Re}\left\{ \operatorname{Tr}\left(\mathbf{R}_p^{\mathrm{H}} \mathbf{R}_{p-1} \mathbf{A}_i \right) \right\} c_i^{\mathrm{R}} + \\ \operatorname{Re}\left\{ \operatorname{Tr}\left(\mathbf{R}_p^{\mathrm{H}} \mathbf{R}_{p-1} j \mathbf{B}_i \right) \right\} c_i^{\mathrm{I}} \end{array} \right] \right], \\[2ex]
\{\hat{c}_2, \hat{c}_3\} &= \arg \underset{\{c_2, c_3\} \in \mathcal{M}}{\operatorname{Max}} \left[\sum_{i=2,3} \left[\begin{array}{l} \operatorname{Re}\left\{ \operatorname{Tr}\left(\mathbf{R}_p^{\mathrm{H}} \mathbf{R}_{p-1} \mathbf{A}_i \right) \right\} c_i^{\mathrm{R}} + \\ \operatorname{Re}\left\{ \operatorname{Tr}\left(\mathbf{R}_p^{\mathrm{H}} \mathbf{R}_{p-1} j \mathbf{B}_i \right) \right\} c_i^{\mathrm{I}} \end{array} \right] \right].
\end{aligned} \tag{5.14}
$$

As shown in (5.14), the proposed differential QO-STBC modulation scheme can be decoded by two joint detections each involving two complex symbols, and the two decision metrics can be computed in parallel. So the proposed DSTM scheme is double–symbol–decodable, just like the original coherent **J4** code. Being double–symbol–decodable, the proposed DSTM scheme has much lower decoding complexity than other unitary DSTM schemes reported in the literature (with the exception of [63]), which generally require much larger joint detection search space dimensions for the same spectral efficiency and antenna number. Complexity and performance comparisons between the various DSTM schemes will be elaborated in Section 5.2.4.

5.2.3.2   Design of constellation set

In this section, we present a pair-wise constellation set $\mathcal{M}$ with good scalability in spectral efficiency that satisfies all three requirements in (5.12) with $v = 0$ ($v = 0$ is chosen because it gives an elegant tractable solution. Our empirical studies with some $v \neq 0$ values do not give better solutions than $v = 0$). The proposed constellation set $\mathcal{M}$ is:

$$\{a_i,b_i\} = \begin{cases} a_i = \sqrt{2}\exp\left[j\left(2k\pi/M\right)\right], \\ b_i = 0, \end{cases} \qquad \forall\, 1 \leq i \leq L/2,$$

$$\{a_i,b_i\} = \begin{cases} a_i = 0, \\ b_i = \sqrt{2}\exp\left[j\left(2(k-L/2)\pi/M+\theta\right)\right], \end{cases} \qquad \forall\, L/2 < i \leq L, \tag{5.15}$$

where $M = L/2$ is an integer, and θ lies in between 0 and $2\pi/M$. Note that in (5.15), a_i belongs to a M-ary PSK constellation set, for $1 \leq i \leq L/2$, while b_i belongs to a rotated M-ary PSK constellation set, for $L/2 < i \leq L$. The design idea behind (5.15) is as follows: By choosing a zero-value for a_i or b_i, the first criterion of (5.12) is met for $v = 0$; while by drawing the other a_i or b_i from the MPSK constellation shown in (5.15), all remaining criteria in (5.12) are met too. The parameter M is related to the spectral efficiency R by $R = 2(\log_2 2M)/N_T$ (since $L = 2M$), hence a DSTM scheme with a wide range of spectral efficiency can be systematically designed from (5.15) by adjusting M. The constellation rotation (CR) angle θ for a quarter of the symbols in $\mathcal{M}$ provides an extra degree of freedom to maximize the diversity and coding gain of the resultant DSTM scheme.

<u>*Theorem 5.1*</u>: For the DSTM constellation set defined in (5.15), the optimum value of the CR angle θ to meet the Performance Optimization Criterion (5.12)(iii) is π/M when M is even, and $\pi/2M$ or $3\pi/2M$ when M is odd.

Proof of Theorem 5.1: The proof is given in four cases, considering different values of i and k, where i and k represent different symbol indexes:

Case 1: $1 \leq i, k \leq L/2$, and $i \neq k$.
Since b_i and b_k are always zero in this case, Δb_{ik} is always zero, but not Δa_{ik} (since $i \neq k$). Hence the determinant value $\left[\left| \Delta a_{ik} + \Delta b_{ik} \right|^2 \times \left| \Delta a_{ik} - \Delta b_{ik} \right|^2 \right]^2$ in (5.12) (iii) can never be zero, and its value is independent of θ. This implies that full diversity is always achieved in this case, and the coding gain optimization does not depend on θ.

Case 2: $L/2 < i, k \leq L$ and $i \neq k$.
The proof is similar to Case 1 with the roles of Δa_{ik} and Δb_{ik} interchanged.

Case 3: $1 \leq i \leq L/2$ and $L/2 < k \leq L$.
For this case, the determinant value $\left[\left| \Delta a_{ik} + \Delta b_{ik} \right|^2 \times \left| \Delta a_{ik} - \Delta b_{ik} \right|^2 \right]^2$ in (5.12) (iii) can be simplified to:

$$
\begin{aligned}
\det &= \frac{1}{16} \left[\begin{array}{l} \left(\left| \exp\left[j\left(2i\pi / M \right) \right] + \exp\left[j\left(2(k - L/2)\pi / M + \theta \right) \right] \right|^2 \right) \times \\ \left(\left| \exp\left[j\left(2i\pi / M \right) \right] - \exp\left[j\left(2(k - L/2)\pi / M + \theta \right) \right] \right|^2 \right) \end{array} \right]^2 \\[2mm]
&= \left[\sin^2 \left(\frac{2\pi i}{M} - \frac{2\pi q}{M} - \theta \right) \right]^2 \qquad 1 \leq i, q \triangleq k - \frac{L}{2} \leq M \\[2mm]
&= \sin^4 \left(\frac{2\pi (l - m)}{M} \right), \qquad\qquad 0 \leq l \triangleq i - q \leq M - 1
\end{aligned}
$$

(5.16)

where i, k, l, q are integers, $M \triangleq L/2$, and $m \triangleq (M\theta)/(2\pi)$ is a real number between 0 and 1 inclusively.

To maximize the minimum determinant value in (5.16), we first consider *even* values of M. As shown in Fig. 5.1(a), the function $\sin[2l\pi/M]$ (solid line) is zero if $l = 0$ or $M/2$, as shown by the *triangular* markers. This results in a zero determinant value in (5.16) and the resultant DSTM will not deliver full diversity. In order to achieve full diversity and maximum coding gain, a right shift m can be introduced to obtain the function $\sin[2(l-m)\pi/M]$ (dashed line) such that its minimum *absolute* values at integer values of l are non-zero and maximized, as shown by the *circular* markers in Fig. 5.1(a). Clearly, this optimum point is reached when $m \triangleq (M\theta)/(2\pi) = 0.5$, which corresponds to $\theta = \pi/M$. Similarly, as indicated by the circular markers in Fig. 5.1(b), the optimum m value for *odd* values of M is 0.25 or 0.75, which corresponds to $\theta = \pi/2M$ or $3\pi/2M$ respectively.

Case 4: $1 \leq k \leq L/2$ and $L/2 < i \leq L$.
The proof is similar to Case 3.

As the determinant value in (5.12)(iii) does not depend on θ for Cases 1 and 2, the optimum θ value is determined by Cases 3 and 4 to be π/M for M even and $\pi/2M$ or $3\pi/2M$ for M odd, hence *Theorem 5.1* is proved.

∎

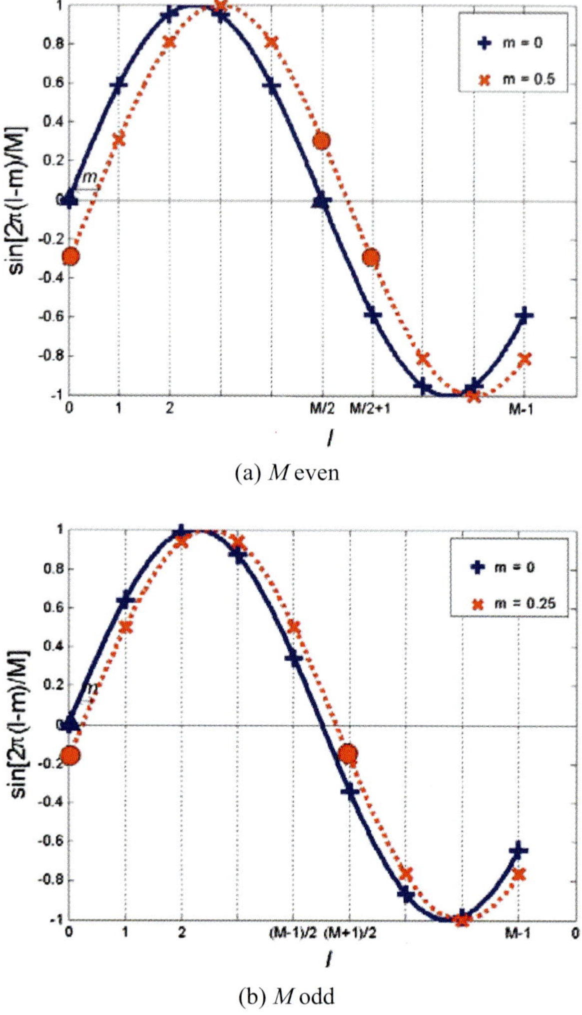

(a) *M* even

(b) *M* odd

Fig. 5.1 Optimization of constellation rotation angle $\theta = 2\pi m / M$ in (5.15)

5.2.4 *Performance comparison*

In Table 5.1, the proposed double–symbol–decodable DSTM with optimized joint constellation set as specified in (5.15) and *Theorem 5.1* is compared with DSTM schemes based on O-STBC [63] and group codes

[62] in terms of coding gain in (5.3) and decoding complexity. We can see that the proposed DSTM provides the highest coding gain at both spectral efficiency values of 1.5bps/Hz and 2 bps/Hz. The proposed DSTM also has a much lower decoding complexity than the group-code-based DSTM in [62], with a decoding search space dimension of 8 at 1.5bps/Hz and 16 at 2 bps/Hz. Although the DSTM based on rate-3/4 O-STBC in [63] has a lower decoding complexity than the proposed scheme at a spectral efficiency of 1.5bps/Hz, it has a lower coding gain. At a spectral efficiency of 2 bps/Hz, the proposed DSTM has a higher coding gain and the same decoding search space dimension as the DSTM based on rate-1/2 O-STBC with 16PSK [63].

Table 5.1 Comparison of coding gains and decoding complexity for DSTM schemes for four transmit antennas

Spectral Efficiency	DSTM Scheme	Constellation	Coding Gain	Decoding Search Space Dimension
1.5	Group Code [62]	64PSK	1.85	64
1.5	**Proposed Double–Symbol– Decodable DSTM**	**(5.15) with** $M = 4,\ \theta = \pi/4$	**2.83**	**8**
1.5	Rate-3/4 O-STBC [63]	QPSK	2.70	4
2	Group Code [62]	256PSK	0.78	256
2	**Proposed Double–Symbol– Decodable DSTM**	**(5.15) with** $M = 8,\ \theta = \pi/8$	**1.17**	**16**
2	Rate-1/2 O-STBC [63]	16PSK	0.31	16

In Fig. 5.2, we compare the block error rate (BLER) performance of the proposed DSTM scheme, obtained through simulations, with that of the DSTM schemes reported in [63,64,76], assuming four transmit and one receive antennas. Note that the DSTM schemes of Cayley [76] and Sp(2) [64] cannot support 2 bps/Hz exactly.

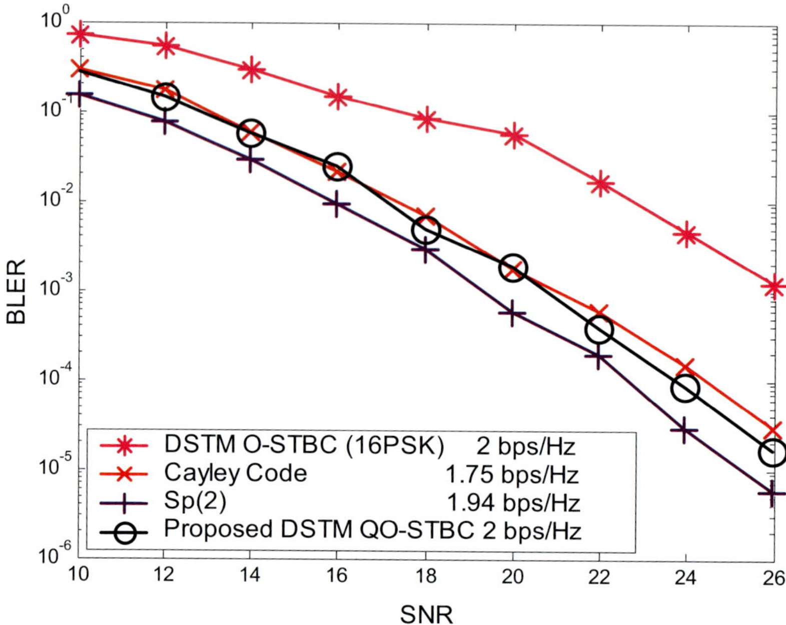

Fig. 5.2 BLER of different DSTM schemes for four transmit and one receive antennas

Compared with the DSTM scheme of Cayley [76] of spectral efficiency 1.75 bps/Hz, with $Q = 7$, $r = 2$, DSTM scheme in [73] based on QO-STBC provides a higher spectral efficiency of 2 bps/Hz with comparable BLER performance. Compared with the DSTM scheme based on Sp(2) [64], with $M = 5$, $N = 3$, of spectral efficiency 1.94 bps/Hz and decoding search space dimension of 225, the DSTM scheme based on QO-STBC performs 1dB worse in BLER but has a much smaller decoding search space dimension of 16 and a slightly higher spectral efficiency. Finally, compared with the rate-1/2 O-STBC [63] DSTM scheme of spectral efficiency 2 bps/Hz, the DSTM scheme based on QO-STBC has a 5 db gain at BLER $= 10^{-3}$ with the same spectral efficiency and decoding search space dimension. Such significant performance gain is possible because the proposed DSTM scheme based on QO-STBC has a higher code rate, hence it can use a constellation with a larger minimum Euclidean distance to support a given spectral efficiency.

5.2.5 Section summary

A unitary double–symbol–decodable DSTM scheme can be constructed from a double–symbol–decodable QO-STBC. The main idea is to force the QO-STBC codeword to be a unitary matrix by using pairwise symbol modulation with a specially designed constellation set. Such unitary double–symbol–decodable DSTM scheme inherits the merits of QO-STBC, hence it achieves a higher code rate and better decoding performance than DSTM scheme based on O-STBC. It is also has lower decoding complexity (smaller decoding search space dimension) than other DSTM schemes not based on O-STBC, with comparable decoding performance.

5.3 Quasi-Unitary DSTM Based on MDC-QOSTBC

5.3.1 Literature review

As discussed in Section 5.2, the first-generation DSTM schemes were designed based on *unitary* matrices. Due to the unitary requirement imposed on the underlying space-time code, unitary DSTMs mostly employ phase shift keying (PSK) constellation. It is well known that PSK constellation becomes less power and spectrum efficient as the constellation size increases. To overcome this limitation, DSTMs based on *quasi-unitary* matrices have been proposed. In [77], a DSTM based on O-STBC with Amplitude-PSK (A-PSK) constellation was first proposed; subsequently in [71,72], DSTMs based on O-STBC with general QAM constellation were designed. They were shown to have better performance than unitary DSTM schemes with PSK constellations. Likewise, a DSTM based on QO-STBC with non-constant matrix norm was proposed in [78].

In this section, we introduce a new quasi-unitary and single–symbol–decodable DSTM scheme [74,75]. This is the first single–symbol–decodable DSTM scheme with higher code rate than O-STBC DSTM. We will show its construction based on the Minimum–Decoding–

Complexity Quasi-Orthogonal Space-Time Block Code (MDC-QOSTBC)[2] presented in Chapter 4. Compared with the unitary double–symbol–decodable DSTM presented earlier in Section 5.2, this quasi-unitary single–symbol–decodable DSTM has even smaller decoding search space dimension, hence lower decoding complexity. Its code design criteria, constellation set design and error probability performance will be evaluated.

5.3.2 *Signal model of quasi-unitary DSTM scheme*

Consider a MIMO channel with N_T transmit antennas and N_R receive antennas. Let $\mathbf{H}_p$ be the $N_R \times N_T$ fading channel gain matrix at a time p. The $(i, k)^{\text{th}}$ element of $\mathbf{H}_p$ is the channel coefficient for the signal path from the kth transmit antenna to the ith receive antenna. Let $\mathbf{X}_p$ be the $N_T \times N_T$ DSTM square codeword transmitted at a time p. Then, the received $N_R \times N_T$ signal matrix $\mathbf{R}_p$ can be written as

$$\mathbf{R}_p = \mathbf{H}_p \mathbf{X}_p + \mathbf{N}_p, \qquad (5.17)$$

where $\mathbf{N}_p$ is the additive white Gaussian channel noise.

At the start of the transmission, a known unitary codeword $\mathbf{X}_0$ is transmitted The codeword $\mathbf{X}_p$ transmitted at a time p is then differentially encoded by

$$\mathbf{X}_p = a_{p-1}^{-1} \mathbf{X}_{p-1} \mathbf{U}_p, \qquad (5.18)$$

where $\mathbf{U}_p$ is a quasi-unitary matrix (i.e. $\mathbf{U}_p \mathbf{U}_p^{H} = a_p^2 \mathbf{I}$) called the *code matrix* that contains information of the transmitted data and a_{p-1} is the square root of the diagonal element of $\mathbf{U}_{p-1} \mathbf{U}_{p-1}^{H}$. To ensure that the average transmission power is maintained constant, it is required that

$$\mathrm{E}\left(\mathbf{X}_p \mathbf{X}_p^{H}\right) = \mathrm{E}\left(\mathbf{U}_p \mathbf{U}_p^{H}\right) = \mathbf{I}, \text{ or } \mathrm{E}\left(a_p^2\right) = 1. \qquad (5.19)$$

---

[2] MDC-QOSTBC is not applicable to the unitary DSTM scheme presented in Section 5.2 because the only solution in this case has a spectral efficiency of 1 bps/Hz. At 1 bps/Hz, a better option is already provided by DSTM based on rate-1 O-STBC with real constellation.

From (5.19), we know that the average value of a_p^2 should be one, but a_p need not to be exactly one at all time (for all p). This is what makes the quasi-unitary design in (5.18) different to the unitary design in (5.5).

If we assume that the channel remains unchanged during two consecutive code periods, i.e. $\mathbf{H}_p = \mathbf{H}_{p-1}$, then the received signal $\mathbf{R}_p$ at a time p can be expressed as

$$\mathbf{R}_p = a_{p-1}^{-1}\mathbf{R}_{p-1}\mathbf{U}_p + \tilde{\mathbf{N}}_p,\tag{5.20}$$

where $\tilde{\mathbf{N}}_p = -a_{p-1}^{-1}\mathbf{N}_{p-1}\mathbf{U}_p + \mathbf{N}_p$. Since $\tilde{\mathbf{N}}_p$ is not independent of a_{p-1} and $\mathbf{U}_p$, the optimal DSTM decoder is a frame-based decoder that performs decoding on the entire sequence of codewords contained in a frame. Clearly it has a very high computational complexity which increases exponentially with the frame length. Hence a near-optimal block-based differential decoder that performs decoding one block/codeword at a time has been proposed in [71] to estimate $\mathbf{U}_p$:

$$\begin{aligned}
\hat{\mathbf{U}}_p &= \arg\underset{\mathbf{U}_p \in \mathcal{U}}{\mathrm{Min}}\left\|\mathbf{R}_p - a_{p-1}^{-1}\mathbf{R}_{p-1}\mathbf{U}_p\right\|^2\\
&= \arg\underset{\mathbf{U}_p \in \mathcal{U}}{\mathrm{Min}}\,\mathrm{Tr}\left(\begin{array}{l}a_{p-1}^{-2}\mathbf{U}_p^{H}\mathbf{R}_{p-1}^{H}\mathbf{R}_{p-1}\mathbf{U}_p - \\ a_{p-1}^{-1}\mathbf{R}_p^{H}\mathbf{R}_{p-1}\mathbf{U}_p - a_{p-1}^{-1}\mathbf{U}_p^{H}\mathbf{R}_{p-1}^{H}\mathbf{R}_p\end{array}\right)\\
&= \arg\underset{\mathbf{U}_p \in \mathcal{U}}{\mathrm{Min}}\,\mathrm{Tr}\left[\begin{array}{l}a_{p-1}^{-2}\mathbf{R}_{p-1}^{H}\mathbf{R}_{p-1}\mathbf{U}_p\mathbf{U}_p^{H} - \\ 2a_{p-1}^{-1}\,\mathrm{Re}\left(\mathbf{R}_p^{H}\mathbf{R}_{p-1}\mathbf{U}_p\right)\end{array}\right],
\end{aligned}\tag{5.21}$$

where $\mathcal{U}$ denotes the set of all possible code matrices, and a_{p-1} can be estimated from the previous decision $\hat{\mathbf{U}}_{p-1}$[3]. Clearly, the near-optimal differential decoder in (5.21) leads to a significant reduction in decoding complexity over the optimal decoder. It has also been shown to have very similar performance as the optimal decoder [71]. Hence we adopt the near-optimal decoder for the proposed single–symbol–decodable DSTM scheme based on MDC-QOSTBC.

---

[3] This may lead to error propagation, but it has been shown in [71,72,78] that the effect of error propagation is small. This will also be confirmed by simulation later in Fig. 5.7 of this chapter.

5.3.3 *Single–symbol–decodable quasi-unitary DSTM*

5.3.3.1 *Quasi-unitary DSTM based on single–symbol–decodable MDC-QOSTBC*

To use the single–symbol–decodable MDC-QOSTBC proposed in Chapter 4 as the code matrix $\mathbf{U}_p$ of a quasi-unitary DSTM scheme, the MDC-QOSTBC codeword $\mathbf{C}$ must have the property that $\mathbf{CC}^{\mathrm{H}} = \gamma\mathbf{I}$ where γ is a constant. For the MDC-QOSTBC constructed from (4.21):

$$\mathbf{CC}^{\mathrm{H}} = \frac{\alpha}{K}\begin{bmatrix} \mathbf{I}_{N_T/2} & \mathbf{0} \\ \mathbf{0} & \mathbf{I}_{N_T/2} \end{bmatrix} + \frac{\beta}{K}\begin{bmatrix} \mathbf{0} & \mathbf{I}_{N_T/2} \\ \mathbf{I}_{N_T/2} & \mathbf{0} \end{bmatrix} \tag{5.22}$$

where

$$\alpha = \sum_{i=1}^{K}\left|c_i\right|^2$$

$$\beta = 2\sum_{i=1}^{K/2}\left(-c_i^{\mathrm{R}}c_i^{\mathrm{I}} + c_{(K/2)+i}^{\mathrm{R}}c_{(K/2)+i}^{\mathrm{I}}\right) \tag{5.23}$$

and its minimum determinant value in (4.31) is:

$$\det = \frac{1}{K}\left[\left(\Delta_i^{\mathrm{R}}\right)^2 - \left(\Delta_i^{\mathrm{I}}\right)^2\right]^{N_T} \tag{5.24}$$

where $\Delta_i = c_i - e_i$, $1 \le i \le 4$, represents the possible error in c_i when the receiver decides erroneously in favor of e_i assuming that c_i was transmitted.

For $\mathbf{C}$ in (5.22) to be quasi-unitary, its α and β in (5.23) must satisfy the conditions that $\beta = 0$ and $\mathrm{E}(\alpha) = K$, where $\mathrm{E}(.)$ is the expectation operator. However, if the code symbols c_i in $\mathbf{C}$ are independent PSK or QAM symbols, then β may not be zero. Hence MDC-QOSTBC with independent PSK or QAM constellation is not quasi-unitary and cannot be used as a DSTM code matrix.

To achieve $\beta = 0$, we know from (5.23) that $c_i^{\mathrm{R}}c_i^{\mathrm{I}}$ must be equal to $c_{(K/2)+i}^{\mathrm{R}}c_{(K/2)+i}^{\mathrm{I}}$. For example, if $K = 4$, we must have $c_1^{\mathrm{R}}c_1^{\mathrm{I}} = c_3^{\mathrm{R}}c_3^{\mathrm{I}}$ and $c_2^{\mathrm{R}}c_2^{\mathrm{I}} = c_4^{\mathrm{R}}c_4^{\mathrm{I}}$. This can be achieved by mapping c_i to a complex constellation symbol $z_k = x_k + jy_k$ such that $x_ky_k = v$ for all k, where v

is a constant. Next, to obtain $E(\alpha) = K$, equation (5.23) suggests that the mapped constellation symbol $z_k = x_k + jy_k$ must satisfy the additional constraint of $E\left(x_k^2 + y_k^2\right) = 1$, assuming that all transmitted symbols have the same power. Finally, to maximize the coding gain, the constellation symbols z_k should also be designed to achieve the maximum possible minimum determinant value in (5.24)[4]. In summary, the collection of constellation symbols $z_k = x_k + jy_k$, $1 \le k \le M$, denoted as constellation set $\mathcal{M}$, for use in a quasi-unitary DSTM based on MDC-QOSTBC, should be designed with the following requirements:

$$\text{(i) Quasi-Unitary Criterion: } x_k y_k = v ;$$

$$\text{(ii) Power Criterion: } E\left(x_k^2 + y_k^2\right) = 1 ; \qquad (5.25)$$

(iii) Performance Criterion:

$$\underset{\arg x_k, y_k}{\text{maximize}} \ \min\left\{\left[\left(\Delta x_{ik}\right)^2 - \left(\Delta y_{ik}\right)^2\right]^{N_T}\right\},$$

where v can be any constant value, and $\Delta x_{ik} = x_i - x_k$, $\Delta y_{ik} = y_i - y_k$ for $1 \le i \ne k \le M$.

The spectral efficiency of the resultant DSTM scheme constructed using an MDC-QOSTBC with code rate R will be $(R \log_2 M)$ bps/Hz. Its decoding decision metric in (5.21) can be reduced to:

---

[4] Since there is no formal design criteria for quasi-unitary DSTM, we borrow the idea from designing unitary DSTM and apply on it. From the results obtained, it seems to be a reasonably good bet.

$$\hat{\mathbf{U}}_p = \underset{\mathbf{U}_p \in \mathcal{U}}{\arg \operatorname{Min}} \operatorname{Tr}\left[a_{p-1}^{-2} \mathbf{R}_{p-1}^{\mathrm{H}} \mathbf{R}_{p-1} \mathbf{U}_p \mathbf{U}_p^{\mathrm{H}} - 2a_{p-1}^{-1} \operatorname{Re}\left(\mathbf{R}_p^{\mathrm{H}} \mathbf{R}_{p-1} \mathbf{U}_p \right) \right]$$

$$= \underset{c_1,\ldots,c_K \in \mathcal{M}}{\arg \operatorname{Min}} \operatorname{Tr}\left[\begin{array}{c} a_{p-1}^{-2} \mathbf{R}_{p-1}^{\mathrm{H}} \mathbf{R}_{p-1} \left(\sum_{i=1}^{K} \frac{|c_i|^2}{K} \mathbf{I}_{N_T} \right) - \\[2ex] 2a_{p-1}^{-1} \operatorname{Re}\left(\mathbf{R}_p^{\mathrm{H}} \mathbf{R}_{p-1} \left(\frac{1}{\sqrt{K}} \sum_{i=1}^{K} \left(c_i^{\mathrm{R}} \mathbf{A}_i + j c_i^{\mathrm{I}} \mathbf{B}_i \right) \right) \right) \end{array} \right]$$

$$= \underset{c_1,\ldots,c_K \in \mathcal{M}}{\arg \operatorname{Min}} \sum_{i=1}^{K} \operatorname{Tr}\left[\begin{array}{c} a_{p-1}^{-2} \left(\frac{|c_i|^2}{K} \right) \mathbf{R}_{p-1}^{\mathrm{H}} \mathbf{R}_{p-1} - \\[2ex] \frac{2a_{p-1}^{-1}}{\sqrt{K}} \operatorname{Re}\left(\mathbf{R}_p^{\mathrm{H}} \mathbf{R}_{p-1} \left(c_i^{\mathrm{R}} \mathbf{A}_i + j c_i^{\mathrm{I}} \mathbf{B}_i \right) \right) \end{array} \right],$$

$$(5.26)$$

where the $\mathbf{A}_i$ and $\mathbf{B}_i$ are the dispersion matrices defined in (4.21). Since every trace term in (5.26) is dependent on only a single c_i symbol, minimizing the sum of all terms is equivalent to minimizing each term seperately. Hence the individual symbols c_i in $\hat{\mathbf{U}}_p$ in (5.26) can be separately detected using the simplified expression below:

$$\hat{c}_i = \underset{c_i \in \mathcal{M}}{\arg \operatorname{Min}} \operatorname{Tr}\left[\begin{array}{c} a_{p-1}^{-2} \left(\frac{|c_i|^2}{K} \right) \mathbf{R}_{p-1}^{\mathrm{H}} \mathbf{R}_{p-1} - \\[2ex] \frac{2a_{p-1}^{-1}}{\sqrt{K}} \operatorname{Re}\left(\mathbf{R}_p^{\mathrm{H}} \mathbf{R}_{p-1} \left(c_i^{\mathrm{R}} \mathbf{A}_i + j c_i^{\mathrm{I}} \mathbf{B}_i \right) \right) \end{array} \right]. \qquad (5.27)$$

As shown in (5.27), the MDC-QOSTBC DSTM scheme is single-symbol decodable as its decoding can be achieved by the joint detection of two real symbols (c_i^{R} and c_i^{I}), or one complex symbol equivalently. This results in the lowest decoding complexity ever achieved for DSTMs not based on O-STBC, including the double–symbol–decodable DSTM scheme presented earlier in Section 5.2.

5.3.3.2 Constellation design

The solution of the three code design criteria in (5.25) will be derived in this section. As shown in Fig. 5.3, the solutions of (x_k, y_k) to (5.25) (ii) can be viewed as points on L concentric circles with radius $r_1, r_2, ..., r_L$, where r_i are subject to the constraint:

$$\sum_{i=1}^{L} r_i^2 = L, \tag{5.28}$$

so as to achieve $E\left(x_k^2 + y_k^2\right) = E(r_k^2) = 1$. Next, the solution loci to (5.25) (i) can be represented as a hyperbola. Hence the intersection points of the hyperbola with the concentric circles represent constellation points that satisfy both (5.25) (i) and (ii). In Fig. 5.3, every circle has at most four intersection points with the hyperbola, illustrated as A_i, B_i, C_i and D_i, where $1 \leq i \leq L$. However, due to their geometrical symmetry, the pair of constellation points A_i and B_i, or C_i and D_i on the same circle will lead to a zero minimum determinant value in (5.24) and the resultant DSTM will not have full diversity. Hence, only the pair of constellation points A_i and C_i, or B_k and D_k (where $i \neq k$), are permissible on each circle, as indicated by the different markers ("x" denotes tha A_i and C_i pair, "o" denotes the B_i and D_i pair) in Fig. 5.3. Finally, to further comply with (5.25) (iii) in order to maximize the minimum determinant value, the (A_i, C_i) constellation pairs and the (B_k, D_k) constellation pairs are used alternately on the adjacent circles.

The above approach enables us to reduce the variables to be optimized in (5.25) (iii) from M pairs of (x_k, y_k) to v and $r_1, r_2, ..., r_L$ where $L = M/2$, since every circle eventually gives two constellation points as the solutions to (5.25) (i)-(iii).

The optimum v and r_i values for the case of four constellation points will now be derived. Referring to Fig. 5.4, we model the points A_1 and C_1 (i.e. points A and C on the first circle) as $A_1 = r_1 \times \exp(j\theta_1)$ and $C_1 = -r_1 \times \exp(j\theta_1)$, and the points B_2 and D_2 (i.e. points B and D on the second circle) as $B_2 = r_2 \times \exp(j\theta_2)$ and $D_2 = -r_2 \times \exp(j\theta_2)$, $0 < \theta_1, \theta_2 < \pi/2$. First we focus on the optimum value for θ_1 corresponding to condition (5.25) (iii):

$$\Delta x_{A_1 C_1} = \text{real}(A_1) - \text{real}(C_1) = 2r_1 \cos(\theta_1),$$
$$\Delta y_{A_1 C_1} = \text{imag}(A_1) - \text{imag}(C_1) = 2r_1 \sin(\theta_1),$$

$$\left[\left(\Delta x_{A_1 C_1} \right)^2 - \left(\Delta y_{A_1 C_1} \right)^2 \right]^{N_T}$$
$$= \left[\left(2r_1 \cos(\theta_1) \right)^2 - \left(2r_1 \sin(\theta_1) \right)^2 \right]^{N_T}$$
$$= \left[4r_1^2 \left[\cos^2(\theta_1) - \sin^2(\theta_1) \right] \right]^{N_T} \tag{5.29}$$
$$= \left[4r_1^2 \cos(2\theta_1) \right]^{N_T}.$$

To maximize (5.29) with respect to θ_1 for even values of N_T, θ_1 has to be 0 or $\pi/2$ (i.e. A_1 and C_1 lie on the x-axis). This corresponds to $v = 0$. Next, consider the optimum value for θ_2:

$$\Delta x_{A_1 B_2} = \text{real}(A_1) - \text{real}(B_2) = r_1 \cos(\theta_1) - r_2 \cos(\theta_2),$$
$$\Delta y_{A_1 B_2} = \text{imag}(A_1) - \text{imag}(C_2) = r_1 \sin(\theta_1) - r_2 \sin(\theta_2),$$

$$\left[\left(\Delta x_{A_1 B_2} \right)^2 - \left(\Delta y_{A_1 B_2} \right)^2 \right]^{N_T}$$
$$= \left[\left(r_1 \cos(\theta_1) - r_2 \cos(\theta_2) \right)^2 - \left(r_1 \sin(\theta_1) - r_2 \sin(\theta_2) \right)^2 \right]^{N_T} \tag{5.30}$$
$$= \left[r_1^2 \cos(2\theta_1) - 2r_1 r_2 \cos(\theta_1 + \theta_2) + r_2^2 \cos(2\theta_2) \right]^{N_T}.$$

To maximize (5.30) with respect to θ_1 and θ_2, $\theta_1 + \theta_2$ has to be $\pi/2$ on condition that A_1 and B_2 lie in the first quadrant (as they are the intersection points of a hyperbola with the concentric circles). So, $\theta_1 = 0$ and $\theta_2 = \pi/2$, or $\theta_1 = \pi/2$ and $\theta_2 = 0$, are two possible solutions. Again they correspond to $v = 0$. Without loss of generality, choose $\theta_1 = 0$ and $\theta_2 = \pi/2$, then equations (5.29) and (5.30) become:

$$\text{(i)} \left[\left(\Delta x_{A_1 C_1} \right)^2 - \left(\Delta y_{A_1 C_1} \right)^2 \right]^{N_T} = \left[4r_1^2 \right]^{N_T},$$
$$\text{(ii)} \left[\left(\Delta x_{A_1 B_2} \right)^2 - \left(\Delta y_{A_1 B_2} \right)^2 \right]^{N_T} = \left[r_1^2 - r_2^2 \right]^{N_T}. \tag{5.31}$$

To further comply with (5.25) (iii), both terms in (5.31) need to be maximized under the constraint (5.28), i.e.

$$\underset{\arg r_1, r_2}{\text{maximize}}\left\{\left[4r_1^2\right]^{N_T}, \left[r_1^2 - r_2^2\right]^{N_T}\right\} \text{ subject to } r_1^2 + r_2^2 = 2. \quad (5.32)$$

The maximization in (5.32) can be carried out by substituting $r_2^2 = 2 - r_1^2$ and setting $\left|4r_1^2\right| = \left|r_1^2 - r_2^2\right|$. The solutions are $r_1 = \sqrt{1/3} = 0.5774$ and $r_2 = \sqrt{5/3} = 1.291$. Due to symmetry, this derivation also applies to A_1 and D_2, or C_1 and B_2.

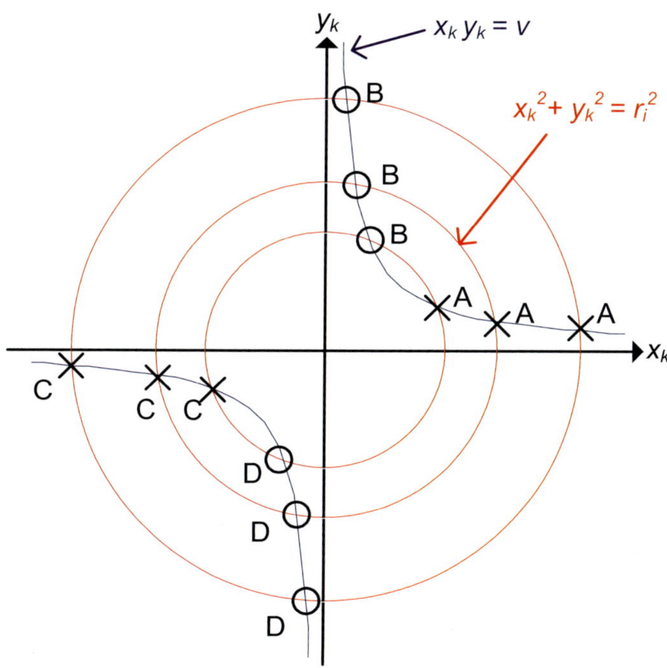

Fig. 5.3 Solution loci of the DSTM constellation design criteria (5.25) (i) and (ii)

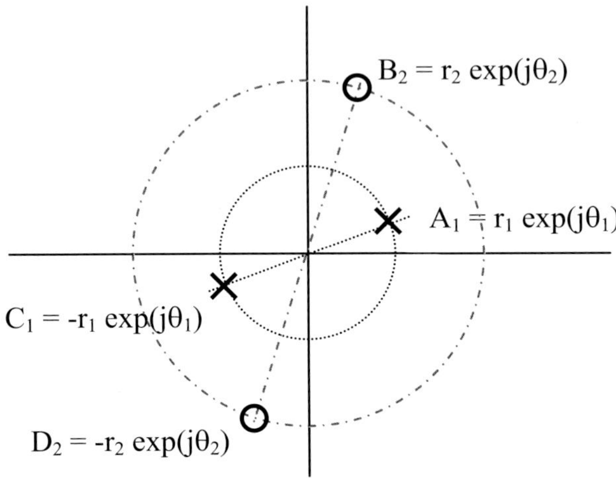

Fig. 5.4 Pre-optimized constellation points for the proposed single–symbol–decodable DSTM based on MDC-QOSTBC

For verification purpose, numerical optimizations have been carried out on computer to obtain the coding gain of the proposed DSTM scheme for different values of v. The numerical solutions obtained for M = 4 and 8 constellation points are shown in Fig. 5.5. They show that the coding gain increases monotonically as v decreases, and $v = 0$ gives the best coding gain for both values of M. This validates the above derivation for the case of $M = 4$ constellation points. For the case of $M = $ 8 constellation points, analytical derivation becomes much more tedious, hence only numerical optimization results are shown in Fig. 5.5.

It should be pointed out that (5.25) (iii) is a max-min optimization problem, whose numerical solutions depend on the initial values and may produce only local maxima. To ensure that the solutions converge, several different initial values have been used in the optimization. Hence it is believed that the coding gain values shown in Fig. 5.5 are close to the global maxima.

In Fig. 5.6, two optimized constellation sets $\mathcal{M}_1$ and $\mathcal{M}_2$ are shown with $M = 4$ and 8 constellation points respectively. Note that these constellation points lie on the x- or y-axis due to $v = 0$, but they are not the standard regular QPSK or Amplitude-PSK constellations. Furthermore, in contrast to coherent MDC-QOSTBC which requires constellation rotation to achieve full diversity [10,32,33], the DSTM constellations shown in Fig. 5.6 already achieve full diversity by design and they should be used *without additional constellation rotation*.

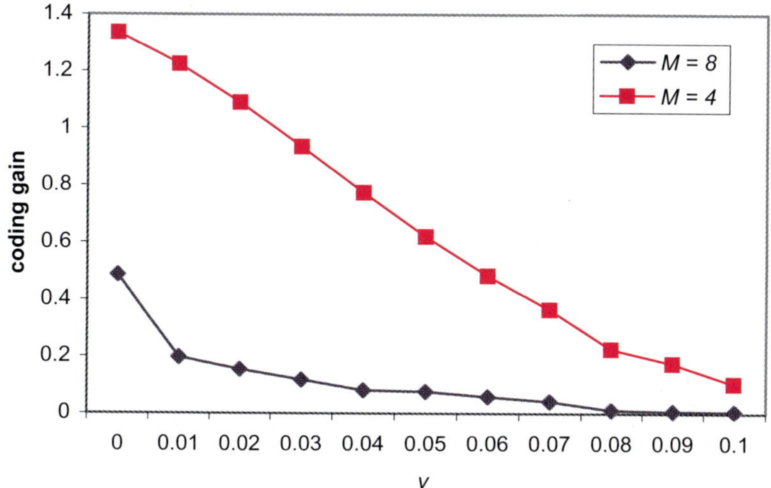

Fig. 5.5 Optimization of coding gain

Quasi-Orthogonal Space-Time Block Code

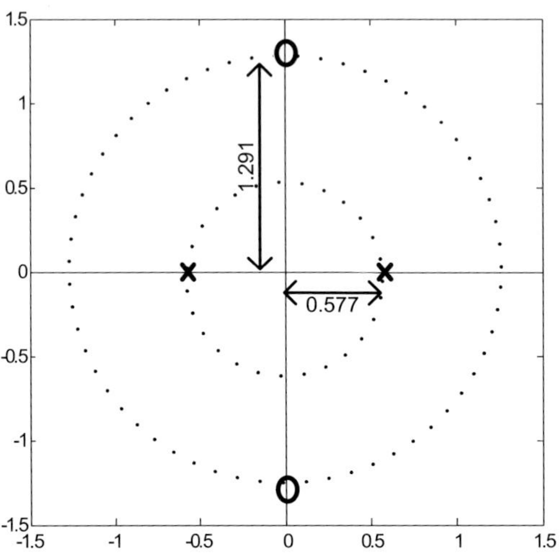

(a) $\mathcal{M}_1$ with $M = 4$ constellation points

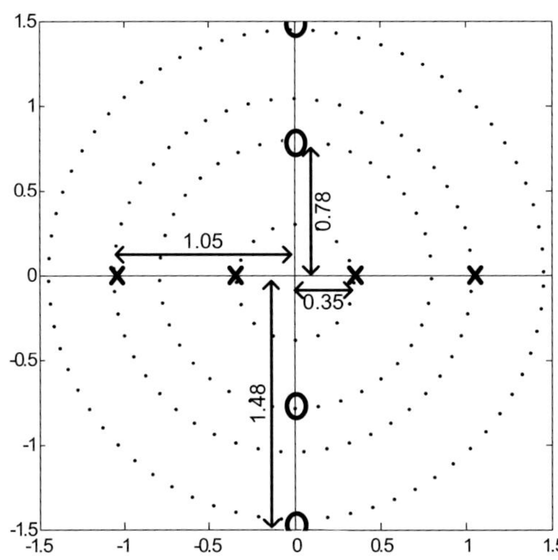

(b) $\mathcal{M}_2$ with $M = 8$ constellation points

Fig. 5.6 Optimized constellation sets for single–symbol–decodable DSTM based on MDC-QOSTBC

5.3.3.3 *Performance comparison*

In the following simulation, the channel is assumed to be flat fading and quasi-static. To facilitate result comparison with [78], every frame from each antenna is assumed to have 132 symbols.

In Fig. 5.7, the BLER performance of the single–symbol–decodable DSTM scheme based on MDC-QOSTBC and the DSTM scheme based on O-STBC [71,72] are compared under common spectral efficiency values. For the case of eight transmit antennas and spectral efficiency of 1.5 bps/Hz, the single–symbol–decodable DSTM scheme based on rate-3/4 MDC-QOSTBC and four-point constellation set $\mathcal{M}_1$ outperforms the DSTM scheme based on rate-1/2 O-STBC and 8QAM constellation, with a 1.5 dB gain at BLER = 10^{-4}. For the case of four transmit antennas and spectral efficiency of 2 bps/Hz, the single–symbol–decodable DSTM scheme based on rate-1 MDC-QOSTBC and four-point constellation set $\mathcal{M}_1$ again outperforms the DSTM scheme based on rate-1/2 O-STBC and 16QAM constellation, with a 3 dB gain at BLER = 10^{-4}. Finally, for four transmit antennas and spectral efficiency of 3 bps/Hz, the DSTM scheme with rate-1 MDC-QOSTBC and eight-point constellation set $\mathcal{M}_2$ performs comparably with the DSTM scheme based on rate-3/4 O-STBC and 16QAM constellation. These results suggest that the single–symbol–decodable DSTM scheme can offer good performance gain over DSTM schemes based on rate-1/2 O-STBC. This is mainly because the single–symbol–decodable DSTM scheme has a higher code rate, hence it can use a constellation with larger Euclidean distance for the same spectral efficiency.

Also included in Fig. 5.7 are the "Genie" BLER results, which are obtained by assuming that the exact values of a_{p-1} are available to the DSTM decoding in (5.27). These "Genie" results are close to the ones obtained by estimating a_{p-1}. This shows that the effects of error propagation due to the estimation of a_{p-1} in (5.27) are negligible, hence we will use the practical (non-Genie) DSTM decoder subsequently in this chapter. Finally, Fig. 5.7 shows that the single–symbol–decodable DSTM scheme based on MDC-QOSTBC has the same BLER slope as the DSTM schemes based on O-STBC. This proves that the single–symbol–decodable DSTM schemes achieve full diversity.

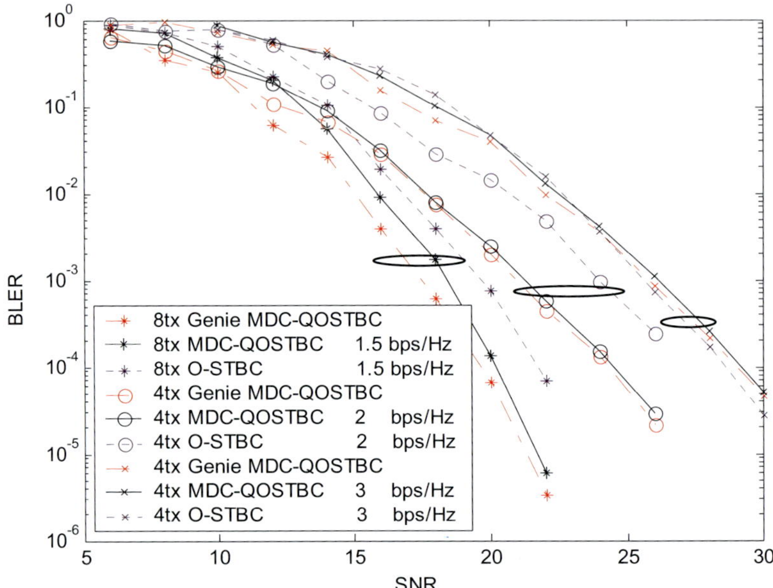

Fig. 5.7 BLER of DSTM schemes based on O-STBC and MDC-QOSTBC

In Fig. 5.8, the BLER performance and decoding complexity of the single–symbol–decodable DSTM scheme for four transmit and one receive antennas with the DSTM schemes based on rate-1 double–symbol–decodable QO-STBC described in Section 5.2, and DSTM based on Sp(2) in [64]. First, the BLER performance of the unitary double–symbol–decodable DSTM scheme from Section 5.2 is compared with this quasi-unitary single–symbol–decodable DSTM scheme in this section. Under a spectral efficiency of 3 bps/Hz, the single–symbol–decodable DSTM scheme based on MDC-QOSTBC performs comparably as the double–symbol–decodable DSTM scheme based on QO-STBC, but the decoding search space dimension of the latter (64 constellation points) is the square of the former (8 constellation points).

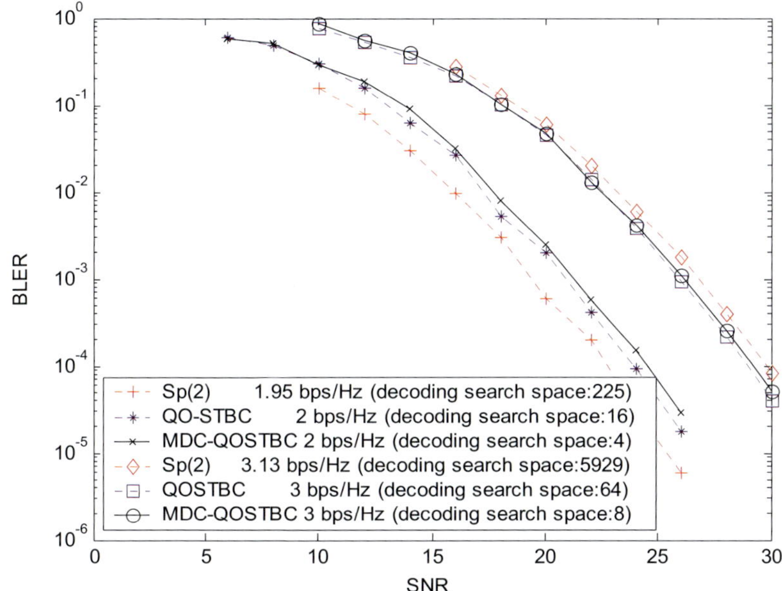

Fig. 5.8 BLER of DSTM schemes for four transmit and one receive antennas

Next, the single–symbol–decodable DSTM scheme with spectral efficiency of 3 bps/Hz has a 0.5dB performance advantage at BLER=10^{-4} over the DSTM scheme based on Sp(2) with spectral efficiency of 3.13 bps/Hz, but the former has a decoding search space dimension of 8 while the later has a much larger decoding search space dimension of 5929. At a lower spectral efficiency, the Sp(2) DSTM scheme with 1.95 bps/Hz and a decoding search space dimension of 225 outperforms the single–symbol–decodable DSTM scheme, but the latter scheme has a slightly higher spectral efficiency of 2 bps/Hz and significantly smaller decoding search space dimension of 4. These results suggest that the single–symbol–decodable quasi-unitary DSTM scheme is able to offer significant reduction in decoding complexity with comparable decoding error probability over the DSTM schemes in [78] and [64].

5.3.3.4   Section summary

A single–symbol–decodable DSTM scheme with full diversity can be constructed using the MDC-QOSTBC. The main idea is to render the MDC-QOSTBC codeword quasi-unitary by customizing its symbol constellation. The single–symbol–decodable DSTM scheme achieves a higher code rate (rate 1 for four transmit and rate 3/4 for eight transmit antennas) than a DSTM scheme based on O-STBC, and much lower decoding complexity than any DSTM schemes not based on O-STBC. Its code rate advantage leads to better error probability performance than DSTM based on rate-1/2 O-STBC, while the decoding complexity advantage is achieved with little or no trade-off in decoding error probability performance than DSTM schemes based on Sp(2) or double–symbol–decodable QO-STBC.

5.4   Chapter Summary

Double–symbol–decodable QO-STBC and single–symbol–decodable MDC-QOSTBC can be used to construct new full-diversity DSTM schemes with high code rates and very low decoding complexity (double–symbol–decodable and single–symbol–decodable DSTM respectively). The double–symbol–decodable DSTM requires pair-wise symbol modulation, while the single–symbol–decodable DSTM requires only symbol–by–symbol modulation. But both of them require customized constellation sets. Both can achieve higher code rate than DSTM schemes based on O-STBC, and lower decoding complexity than non-O-STBC DSTM schemes with only slight trade-off in decoding performance.

Chapter 6

Rate, Complexity and Diversity Trade-Off in QO-STBC

In the earlier chapters, we have focused on reducing the ML decoding complexity of coherent and non-coherent QO-STBC. In this chapter, we will investigate the tradeoff effects between code rate, decoding complexity and diversity level in QO-STBC. We will show that QO-STBC can achieve a higher code rate given a higher decoding complexity, or a higher diversity level given a lower code rate.

6.1 QO-STBC with Rate ≤ 1

6.1.1 Introduction

In practice, a full-rate code is often desirable in order to avoid the need for rate-matching [79]. Full-rate STBC with one receive antenna is also optimal from the information capacity perspective [7]. Although O-STBC has low encoder/decoder complexity, it cannot achieve full rate for more than two transmit antennas, while MDC-QOSTBC cannot achieve full rate for more than four transmit antennas. Hence it is of interest to design new QO-STBC with both full code rate and low decoding complexity for transmit antenna number greater than four.

In [50], a class of full-rate QO-STBC, in which the symbols can always be separated into two groups for any number of transmit antennas, has been proposed. This code requires joint detection of half

141

of the transmitted symbols. In [80,81], a rate-one double–symbol–decodable Co-ordinate Interleaved Orthogonal Design (CIOD) code for up to eight transmit antennas is reported. This code has a lower ML decoding complexity than the codes of [50] because its symbols can be separated into four groups. However, its code design generally cannot be extended to more than eight transmit antennas.

In this section, we apply the code construction framework in *Construction 4.1* in Chapter 4 of this monograph to form a new class of QO-STBC, called Four-Group QO-STBC (4Gp-QOSTBC), which supports full code rate for arbitrary number of transmit antennas [82]. The 4Gp-QOSTBC has the property that the symbols received can always be separated into four groups for independent ML decoding. Therefore it requires only half the joint detection symbols per group than the two-group codes in [9,31,50]. The 4Gp-QOSTBC also has other advantages which will be discussed later in this section.

6.1.2 *Full-rate 4Gp-QOSTBC*

6.1.2.1 *Construction of 4Gp-QOSTBC*

Refering to the definition of QO-STBC in *Definition 3.1*, this class of QO-STBC has the property that $G = 4$. This means that after matched filtering, the received symbols can be separated into four independent groups, and the ML decoder only needs to jointly detect $K / 4$ complex symbols in each group.

The 4Gp-QOSTBC can be systematically generated using the following construction rules:

<u>*Construction 6.1*</u>: Given a 4Gp-QOSTBC for N_t transmit antennas, with code length T, and K sets of dispersion matrices denoted as $\{\tilde{\mathbf{A}}_q, \tilde{\mathbf{B}}_q\}$, $1 \leq q \leq K$, a new 4Gp-QOSTBC with code length $2T$ for $2N_t$ transmit antennas, which consists of $2K$ sets of dispersion matrices denoted as $\{\mathbf{A}_q, \mathbf{B}_q\}$, $1 \leq q \leq 2K$, can be constructed using the following four mapping rules:

$$\text{Rule \#1: } \mathbf{A}_{2q-1} = \begin{bmatrix} \tilde{\mathbf{A}}_q & \mathbf{0} \\ \mathbf{0} & \tilde{\mathbf{A}}_q \end{bmatrix}; \quad \text{Rule \#3: } \mathbf{A}_{2q} = \begin{bmatrix} j\tilde{\mathbf{B}}_q & \mathbf{0} \\ \mathbf{0} & j\tilde{\mathbf{B}}_q \end{bmatrix};$$

$$1 \le q \le K$$

$$\text{Rule \#2: } \mathbf{B}_{2q-1} = \begin{bmatrix} \mathbf{0} & j\tilde{\mathbf{A}}_q \\ j\tilde{\mathbf{A}}_q & \mathbf{0} \end{bmatrix}; \quad \text{Rule \#4: } \mathbf{B}_{2q} = \begin{bmatrix} \mathbf{0} & \tilde{\mathbf{B}}_q \\ \tilde{\mathbf{B}}_q & \mathbf{0} \end{bmatrix}.$$

$$(6.1)$$

Proof of Construction 6.1: This construction is in fact the same as *Construction 4.1*, except that the lower order $\tilde{\mathbf{A}}_q, \tilde{\mathbf{B}}_q$ in *Construction 6.1* are 4Gp-QOSTBC while those in *Construction 4.1* are O-STBC. Hence their proof is the same too. ∎

Theorem 6.1: Given a 4Gp-QOSTBC for N_t transmit antennas, associated with dispersion matrices $\{ \tilde{\mathbf{A}}_q, \tilde{\mathbf{B}}_q \}$, $1 \le q \le K$, satisfies the MSD constraint in (3.6), a new 4Gp-QOSTBC for $2N_t$ transmit antennas, associated with dispersion matrices $\{\mathbf{A}_q, \mathbf{B}_q\}$, $1 \le q \le 2K$, satisfies the MSD constraint as well.

Proof of Theorem 6.1: This can be proved as follow. For example, we first examine the dispersion matrices generated by Rule 1 in (6.1):

$$\mathbf{A}_{2q-1}^H \mathbf{A}_{2q-1} = \begin{bmatrix} \tilde{\mathbf{A}}_q & \mathbf{0} \\ \mathbf{0} & \tilde{\mathbf{A}}_q \end{bmatrix}^H \begin{bmatrix} \tilde{\mathbf{A}}_q & \mathbf{0} \\ \mathbf{0} & \tilde{\mathbf{A}}_q \end{bmatrix}$$

$$= \begin{bmatrix} \tilde{\mathbf{A}}_q^H \tilde{\mathbf{A}}_q & \mathbf{0} \\ \mathbf{0} & \tilde{\mathbf{A}}_q^H \tilde{\mathbf{A}}_q \end{bmatrix}$$

$$= (T/K)\mathbf{I}_{2N_t} \qquad \text{if } \tilde{\mathbf{A}}_q^H \tilde{\mathbf{A}}_q = (T/K)\mathbf{I}_{N_t}$$

Hence, if the lower-order dispersion matrices satisfy the MSD constraint, the higher-order dispersion matrices generated by Rule 1 will satisfy the MSD constraint as well. Likewise it can be shown for Rule 2 to 4, hence *Theorem 6.1* is proved. ∎

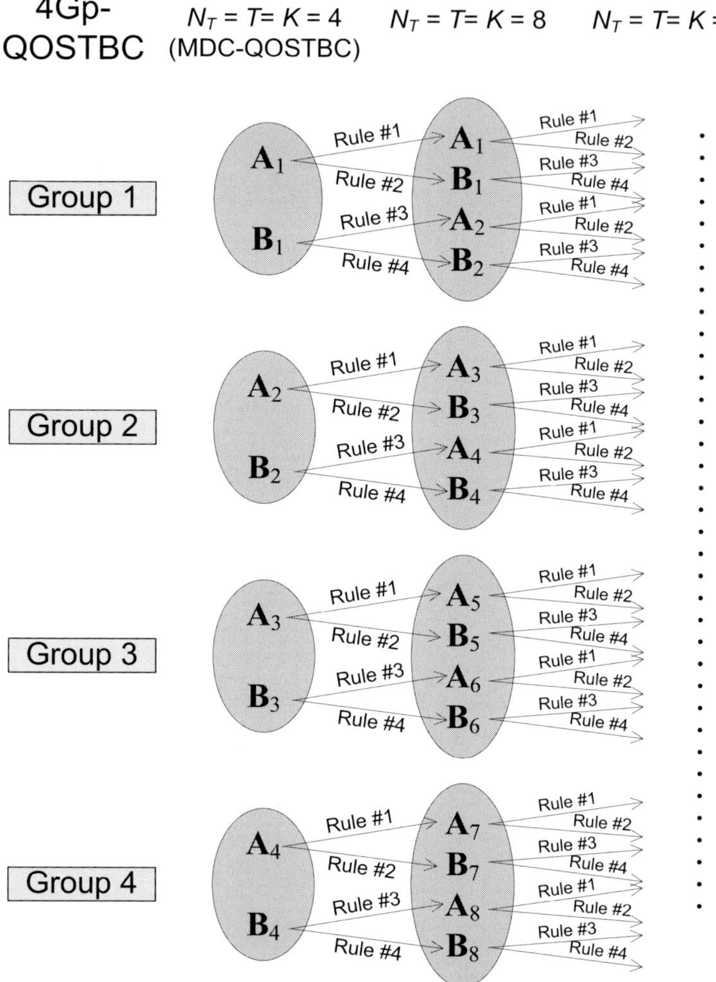

Fig. 6.1 Construction of Full-Rate 4Gp-QOSTBC

The recursive construction of 4Gp-QOSTBC described in (6.1) may start with the MDC-QOSTBC for four transmit antennas presented in Chapter 4 of this monograph and [55,56]. This is because a MDC-QOSTBC is the smallest code that meets the definition of a 4Gp-

QOSTBC. As shown in Fig. 6.1, by using the four mapping rules in *Construction 6.1*, we can generate a 4Gp-QOSTBC for eight transmit antennas, which can in turn generate a 4Gp-QOSTBC for sixteen transmit antennas, and so on. In addition, since a MDC-QOSTBC satisfies the MSD constraint , so will the resultant 4Gp-QOSTBC as a result of *Theorem 6.1*.

A 4Gp-QOSTBC for eight transmit antennas, generated from MDC-QOSTBC, is shown in (6.2):

$$
\begin{bmatrix}
c_1^R + jc_3^R & c_5^R + jc_7^R & -c_2^R + jc_4^R & -c_6^R + jc_8^R & -c_1^I - jc_3^I & -c_5^I - jc_7^I & -c_2^I + jc_4^I & -c_6^I + jc_8^I \\
-c_5^R + jc_7^R & c_1^R - jc_3^R & c_6^R + jc_8^R & -c_2^R - jc_4^R & c_5^I - jc_7^I & -c_1^I + jc_3^I & c_6^I + jc_8^I & -c_2^I - jc_4^I \\
-c_2^R + jc_4^R & -c_6^R + jc_8^R & c_1^R + jc_3^R & c_5^R + jc_7^R & -c_2^I + jc_4^I & -c_6^I + jc_8^I & -c_1^I - jc_3^I & -c_5^I - jc_7^I \\
c_6^R + jc_8^R & -c_2^R - jc_4^R & -c_5^R + jc_7^R & c_1^R - jc_3^R & c_6^I + jc_8^I & -c_2^I - jc_4^I & c_5^I - jc_7^I & -c_1^I + jc_3^I \\
-c_1^I - jc_3^I & -c_5^I - jc_7^I & -c_2^I + jc_4^I & -c_6^I + jc_8^I & c_1^R + jc_3^R & c_5^R + jc_7^R & -c_2^R + jc_4^R & -c_6^R + jc_8^R \\
c_5^I - jc_7^I & -c_1^I + jc_3^I & c_6^I + jc_8^I & -c_2^I - jc_4^I & -c_5^R + jc_7^R & c_1^R - jc_3^R & c_6^R + jc_8^R & -c_2^R - jc_4^R \\
-c_2^I + jc_4^I & -c_6^I + jc_8^I & -c_1^I - jc_3^I & -c_5^I - jc_7^I & -c_2^R + jc_4^R & -c_6^R + jc_8^R & c_1^R + jc_3^R & c_5^R + jc_7^R \\
c_6^I + jc_8^I & -c_2^I - jc_4^I & c_5^I - jc_7^I & -c_1^I + jc_3^I & c_6^R + jc_8^R & -c_2^R - jc_4^R & -c_5^R + jc_7^R & c_1^R - jc_3^R
\end{bmatrix}
. \tag{6.2}
$$

It can be easily shown that the 4Gp-QOSTBC in (6.2) can separate its received symbols into four independent groups by simple linear match filtering. These four symbol groups are $\{c_1, c_2\}$, $\{c_3, c_4\}$, $\{c_5, c_6\}$, and $\{c_7, c_8\}$. Since the code satisfies the MSD constraint, it guarantees full diversity for single (real) symbol error events. Constellation rotation technique can then be used to enable it to achieve full diversity for any other symbol error events.

6.1.2.2 *Code rate of 4Gp-QOSTBC*

By its definition, the higher-order 4Gp-QOSTBC defined in *Theorem 6.1* has code rate $= 2K / 2T = K / T$, which is the same as the code rate of the lower-order 4Gp-QOSTBC that is used to construct it.

<u>*Theorem 6.2*</u>: The maximum achievable code rate of a 4Gp-QOSTBC is 1 for any number of transmint antennas ≥ 4.

<u>*Proof of Theorem 6.2*</u>: It has been shown in Chapter 4, [56], and [57] that the maximum achievable code rate of an MDC-QOSTBC for four transmit antennas is one. Since the higher-order 4Gp-QOSTBC in (6.1) has the same code rate as the lower-order 4Gp-QOSTBC used to generate it, and MDC-QOSTBC is one such choice, we can conclude that the maximum achievable code rate of a 4Gp-QOSTBC is one. The fact that a full-rate MDC-QOSTBC for four transmit antennas exists implies that a rate-1 4Gp-QOSTBC exists for any number of antennas greater than four.   ∎

In addition, if the number of transmit antennas is a power of two, the 4Gp-QOSTBC constructed in (6.1) is a square design, hence it is delay optimal, i.e. the code length is minimal and equal to the number of transmit antennas [17]. It is because MDC-QOSTBC for four transmit antennas is a square design; hence all the 4Gp-QOSTBC constructed by *Construction 6.1* using square MDC-QOSTBC will be square too, if the number of transmit antennas is a power of two, i.e. 8, 16, … and so on. For 4Gp-QOSTBC to support a number of transmit antennas not a power of two, the method described in Section 4.3.3.1 can be applied.

6.1.2.3   *Decoding performance*

As shown in [82], constellation rotation (CR) can be applied to optimize the decoding performance of the 4Gp-QOSTBC codes.[5] In Fig. 6.2, the minimum determinant value of the codeword distance matrix is plotted as a function of the constellation rotation angle of 4-QAM constellation symbols c_1, c_3, c_5 and c_7 (denoted as Angle 1) and symbols c_2, c_4, c_6 and c_8 (denoted as Angle 2). The optimum constellation rotation angle is found through computer search to be 7^0 for the Angle 1 and 23^0 for Angle 2.

---

[5] For 4Gp-QOSTBC, CR is a special case of GCLT (Section 3.4) because 4Gp-QOSTBC is designed to have the real and imaginary parts of a symbol in the same group. Hence CR will not increases the decoding complexity of 4Gp-QOSTBC.

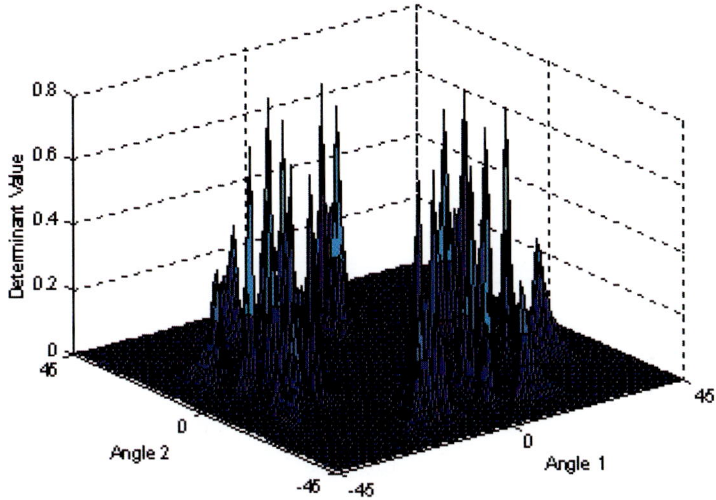

Fig. 6.2 Improving 4Gp-QOSTBC with constellation rotation

In Table 6.1 the decoding complexity and diversity product of various rate-1 QO-STBCs are compared. It shows that the number of symbols required for joint detection by 4Gp-QOSTBC and CIOD [80,81] is half that required by the QO-STBC in [9,31,50]. For the case of five transmit antennas, the diversity products of QO-STBC and 4Gp-QOSTBC are close, and much larger than that of CIOD. For the case of eight transmit antennas, 4Gp-QOSTBC and CIOD has the similar diversity product and we will next evaluate their BER performance.

Table 6.1 Comparisons of decoding complexity and diversity product for various rate-1 QO-STBCs

STBC with CR	Number of real symbols for joint detection	Diversity product	
		5 tx antennas	8 tx antennas
QO-STBC [9,31,50]	8	0.2712	0.2187
4Gp-QOSTBC	4	0.2653	0.1727
CIOD [80,81]	4	0.1790	0.1682

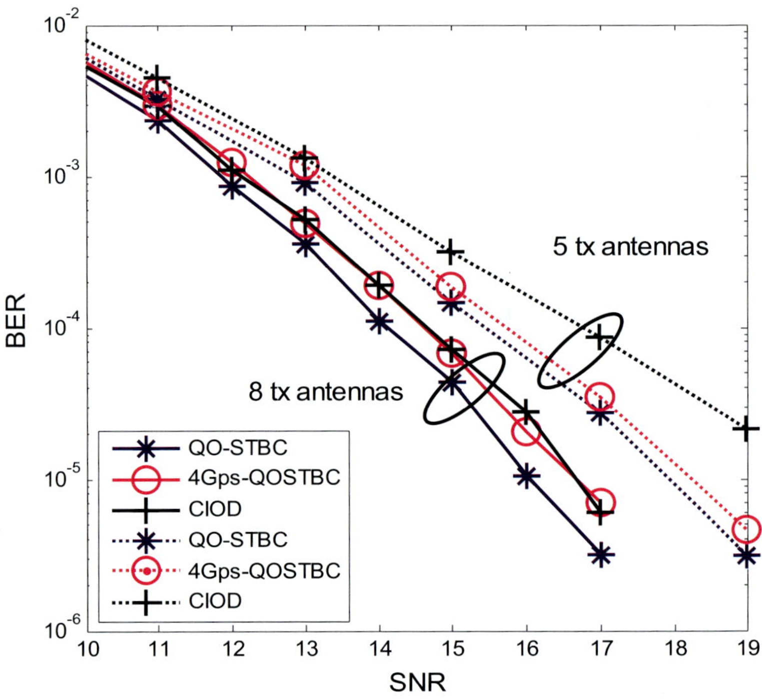

Fig. 6.3 BER of full-rate QO-STBCs for five and eight transmit antennas with 4-QAM constellation

The bit error rate (BER) performance, obtained from simulations, of full-rate 4Gp-QOSTBC for eight and five[6] transmit antennas with 4-QAM constellation are shown in Fig. 6.3. For eight transmit antennas, 4Gp-QOSTBC can be seen to perform approximately 0.5 dB worse at a BER of 10^{-4} when compared with the QO-STBC with two groups from [9,31,50]. However for 4Gp-QOSTBC , the receiver only needs to jointly decode four real symbols. This is half the number of joint detection symbols required in the case of QO-STBC. Furthermore, 4Gp-QOSTBC has the same decoding performance as CIOD,[7] but without the high peak-to-average power ratio problem as there is no zero in the transmission matrix as shown in (6.2). For the case of five transmit antennas, the BER disadvantage of between 4Gp-QOSTBC from the QO-STBC with two groups is smaller, but the former still retains the same complexity advantage. However, compared with CIOD, 4Gp-QOSTBC performs about 1.5dB better at BER 10^{-5}, while their decoding complexity are the same.

6.1.3　Rate–complexity–diversity tradeoff

The four mapping rules in *Construction 6.1* can be extended to QO-STBC with more groups, such that the number of symbols required for ML joint detection is reduced. For example, starting with the rate-3/4 MDC-QOSTBC for eight transmit antennas presented in Section 4.3 that has 6 groups of quasi-orthogonal symbols, a class of rate-3/4 6Gp-QOSTBC for sixteen transmit antennas can be constructed. The same can be said for a rate-1/2 8Gp-QOSTBC, as shown in Table 6.2.

Clearly, a tradeoff between the code rate and decoding complexity in QO-STBC can be seen. For example, for eight transmit antennas, a code rate of 1 can be achieved if a joint detection of four real symbols is allowed However, if one desires only to jointly decode two real symbols, then the code rate that can be achieved is only ¾; this is further reduced

---

[6] The codeword of 4Gp-QOSTBC for five transmit antennas is obtained by removing the last three columns of the transmission matrix for eight transmit antennas.

[7] For CIOD the simulation has been carried out by having the symbols c_1, c_3, c_5, c_7 rotated with 38^0; while symbols c_2, c_4, c_6, c_8 rotated with 23^0.

to rate 1/2, if one only allows linear detection. In general, Table 6.2 shows that every doubling in the number of quasi-orthogonal groups is accompanied by a reduction of 1/4 in the code rate of a QO-STBC. Conversely, one could also expect that every increment of 1/4 in code rate will halve the number of quasi-orthogonal groups in the QO-STBC. This relationship interestingly coincides with a subsequent finding in Section 6.2 that a rate 5/4 QO-STBC with 2 groups exists.

Table 6.2 Properties of QO-STBC constructed using *Construction 6.1*

No. of Gps	Rate	2 Tx	4 Tx	8 Tx	16 Tx
			Diversity increases in this direction → **Complexity increases in this direction →**		
4	1	O-STBC gp size: 1	MDC-QOSTBC gp size: 2	4Gp-QOSTBC gp size: 4	4Gp-QOSTBC gp size: 8
6	3/4	N.A.	O-STBC gp size: 1	MDC-QOSTBC gp size: 2	6Gp-QOSTBC gp size: 4
8	1/2	N.A.	N.A.	O-STBC gp size: 1	MDC-QOSTBC gp size: 2

Rate increases in this direction →
Complexity increases in this direction →

"gp size": number of real symbols per group

6.1.4 Section summary

Four dispersion matrix mapping rules have been designed to construct a class of full-rate QO-STBC, called 4Gp-QOSTBC, in which the received symbols can always be separated into four quasi-orthogonal groups for independent decoding. 4Gp-QOSTBC results in halving of the number of symbols required for joint detection as compared to the existing full-rate 2-group QO-STBC, and it can be generated for arbitrary number of transmit antennas. Essentially 4Gp-QOSTBC fills the research gap (in terms of decoding complexity reduction) left behind by full-rate O-STBC, full-rate MDC-QOSTBC, the existing full-rate 2-group QO-STBC and CIOD. The proposed code construction rules guarantee maximal symbol-wise diversity, hence 4Gp-QOSTBC can always achieve full diversity via simple constellation rotation. Simulation results show that the full-rate 4Gp-QOSTBC performs as well as the existing full-rate 2-group QO-STBC too. To further trade code rate for decoding complexity, the 4Gp-QOSTBC construction rules can be extended to generate 6Gp-QOSTBC and 8Gp-QOSTBC, with successive doubling in the number of quasi-orthogonal groups and successive reduction of 1/4 in code rate.

6.2   QO-STBC with Rate > 1

6.2.1   Introduction

The tradeoff between code rate and decoding complexity in QO-STBC has been demonstrated in the previous section for code rate up to one. The observed rate-complexity relationship suggests that the maximum code rate of QO-STBC may be able to exceed 1 if the decoding complexity is allowed to increase. In this section, we present the results of a search for such "high-rate QO-STBC" with code rate greater than one [83,84]. To our knowledge, this is the first such attempt ever conducted on QO-STBC.

6.2.2   Code search methodology

The following parameters will be used for generating the dispersion matrices of the high-rate QO-STBC:

- *Code length*: The code length, P, of the QO-STBC is assumed be equal to the number of transmit antennas, i.e. $P = N_t$ in this chapter. So the codeword is a square matrix.
- *Matrix Entries*: The entries of the dispersion matrix are set to be $\{0, \pm1, \pm j\}$, as commonly adopted in the literature.
- *Matrix Rank*: The rank of a dispersion matrix is related to the symbol-wise diversity of the STBC. A dispersion matrix with rank equal to N_t is said to achieve the maximal symbol-wise diversity [11,13]. A code with dispersion matrices of rank 2 can never provide transmit diversity greater than 2.
- *Matrix Weight*: This parameter refers to the number of non-zero entries in every row of a dispersion matrix.
- *Number of Group*: Th number of orthogonal groups in a QO-STBC. This parameter is as defined in Section 3.1. For QO-STBC, G is required to be larger than 1 ($G = 1$ gives a fully non-orthogonal STBC).

The code search methodology works as follows: a set of "seed matrices" is first constructed as potential QO-STBC dispersion matrices, then a search among these matrices is carried out to look for valid QO-STBC dispersion matrices. In order to increase the chance of finding QO-STBC with rate greater than 1, it is desired to have as many seed matrices as possible. Hence seed matrices with complex number entries as well as matrices with weights 1 and 2 are considered.

The code search procedure can be broadly formulated into three steps:

(a) Generate a series of N seed matrices with desired parameters. Consider the example of $P = N_t = 4$, rank 4 and weight 2. To generate matrices with these parameters, one starts with the sixteen 2×2 complex Hadamard matrices with entries $\{0, \pm1, \pm j\}$ shown in Fig. 6.4, and use them as the **P** and **Q** sub-matrices in the sixteen 4×4 matrices shown in Fig. 6.5. By doing so, $N = 16^3 = 4096$ matrices of size 4×4, rank 4 and weight 2 can be generated.

(b) From the N seed matrices obtained in step (a), identify and retain those that can be grouped into G groups according to the QOC in (3.20). However this is an NP-complete problem because each of the N matrices can be either in one of the G groups, or not in any group at all, leading to $(G+1)^N$ possible combinations. Hence an efficient algorithm is required to expedite this search/grouping process. A novel method based on graph theory is described in Section 6.2.3 in this chapter to accomplish this task.

(c) From the quasi-orthognal seed matrices obtained in step (b), identify those that give an equivalent channel matrix of rank > $2P$. These are the dispersion matrices of QO-STBC with a code rate greater than one, relationship between the rank of the equivalent channel matrix and the code rate of a QO-STBC will be explained in the following: Assume that M dispersion matrices, grouped into G groups, are found in step (b). The objective of step (c) is to check the rank, $\mathcal{R}$, of the equivalent channel matrix **H** formed by these M dispersion matrices. $\mathcal{R}$ represents the number of distinct real symbols that can be carried by the equivalent channel **H**. If $\mathcal{R} < M$, it implies that $M - \mathcal{R}$ dispersion matrices are linearly dependent on the $\mathcal{R}$ independent dispersion matrices. So only $\mathcal{R}$ out of the M dispersion matrices can be used to form a QO-STBC with a resultant code rate of $\mathcal{R}$ $/(2P)$. On the other hand, $\mathcal{R} = M$ is the maximum achievable value for $\mathcal{R}$. In order to achieve a code rate greater than one, it is required that $\mathcal{R} > 2P$.

$$\begin{bmatrix} 1 & 1 \\ 1 & -1 \end{bmatrix} \quad \begin{bmatrix} 1 & 1 \\ -1 & 1 \end{bmatrix} \quad \begin{bmatrix} 1 & -1 \\ 1 & 1 \end{bmatrix} \quad \begin{bmatrix} -1 & 1 \\ 1 & 1 \end{bmatrix}$$

$$\begin{bmatrix} 1 & 1 \\ j & -j \end{bmatrix} \quad \begin{bmatrix} 1 & 1 \\ -j & j \end{bmatrix} \quad \begin{bmatrix} 1 & -1 \\ j & j \end{bmatrix} \quad \begin{bmatrix} -1 & 1 \\ j & j \end{bmatrix}$$

$$\begin{bmatrix} 1 & j \\ 1 & -j \end{bmatrix} \quad \begin{bmatrix} 1 & j \\ -1 & j \end{bmatrix} \quad \begin{bmatrix} 1 & -j \\ 1 & j \end{bmatrix} \quad \begin{bmatrix} -1 & j \\ 1 & j \end{bmatrix}$$

$$\begin{bmatrix} 1 & j \\ j & 1 \end{bmatrix} \quad \begin{bmatrix} 1 & -j \\ j & -1 \end{bmatrix} \quad \begin{bmatrix} 1 & -j \\ -j & 1 \end{bmatrix} \quad \begin{bmatrix} 1 & j \\ -j & -1 \end{bmatrix}$$

Fig. 6.4 Complex Hadamard matrices of weight two

$$\begin{bmatrix} \mathbf{P} & 0 \\ 0 & \mathbf{Q} \end{bmatrix} \quad \begin{bmatrix} \mathbf{P} & 0 \\ 0 & -\mathbf{Q} \end{bmatrix} \quad \begin{bmatrix} \mathbf{P} & 0 \\ 0 & j\mathbf{Q} \end{bmatrix} \quad \begin{bmatrix} \mathbf{P} & 0 \\ 0 & -j\mathbf{Q} \end{bmatrix}$$

$$\begin{bmatrix} j\mathbf{P} & 0 \\ 0 & j\mathbf{Q} \end{bmatrix} \quad \begin{bmatrix} j\mathbf{P} & 0 \\ 0 & -j\mathbf{Q} \end{bmatrix} \quad \begin{bmatrix} j\mathbf{P} & 0 \\ 0 & -\mathbf{Q} \end{bmatrix} \quad \begin{bmatrix} j\mathbf{P} & 0 \\ 0 & \mathbf{Q} \end{bmatrix}$$

$$\begin{bmatrix} 0 & \mathbf{P} \\ \mathbf{Q} & 0 \end{bmatrix} \quad \begin{bmatrix} 0 & \mathbf{P} \\ -\mathbf{Q} & 0 \end{bmatrix} \quad \begin{bmatrix} 0 & \mathbf{P} \\ j\mathbf{Q} & 0 \end{bmatrix} \quad \begin{bmatrix} 0 & \mathbf{P} \\ -j\mathbf{Q} & 0 \end{bmatrix}$$

$$\begin{bmatrix} 0 & j\mathbf{P} \\ j\mathbf{Q} & 0 \end{bmatrix} \quad \begin{bmatrix} 0 & j\mathbf{P} \\ -j\mathbf{Q} & 0 \end{bmatrix} \quad \begin{bmatrix} 0 & j\mathbf{P} \\ -\mathbf{Q} & 0 \end{bmatrix} \quad \begin{bmatrix} 0 & j\mathbf{P} \\ \mathbf{Q} & 0 \end{bmatrix}$$

Fig. 6.5 Patterns to generate matrices of rank four and weight two

6.2.3   *Graph modelling and modified depth first search for implementing step (b)*

Now an efficient graph-based technique is described in the following to identify the matrices from those found in step (a) that can be grouped into G groups according to the QOC in (3.20). The quasi-orthogonal relationship between the set of $N = 4096$ matrices, that are obtained in the example earlier, can be visualized in Fig. 6.6, which indicates that the matrix index u on the x-axis and the matrix index v on the y-axis for all $1 \leq u, v \leq N$, and marks a dark pixel at (u, v) point if the corresponding $(\mathbf{A}_u, \mathbf{A}_v)$ matrices satisfy the QOC. A close examination of Fig. 6.6 shows that every matrix satisfies the QOC with another 56 matrices. So if Fig. 6.6 is viewed as a matrix, it is a sparse matrix in which only $56/4096 = 1.37\%$ of the matrix entries are non-zero (the general definition of sparse matrix requires its non-zero entries to be less than 10% [85]).

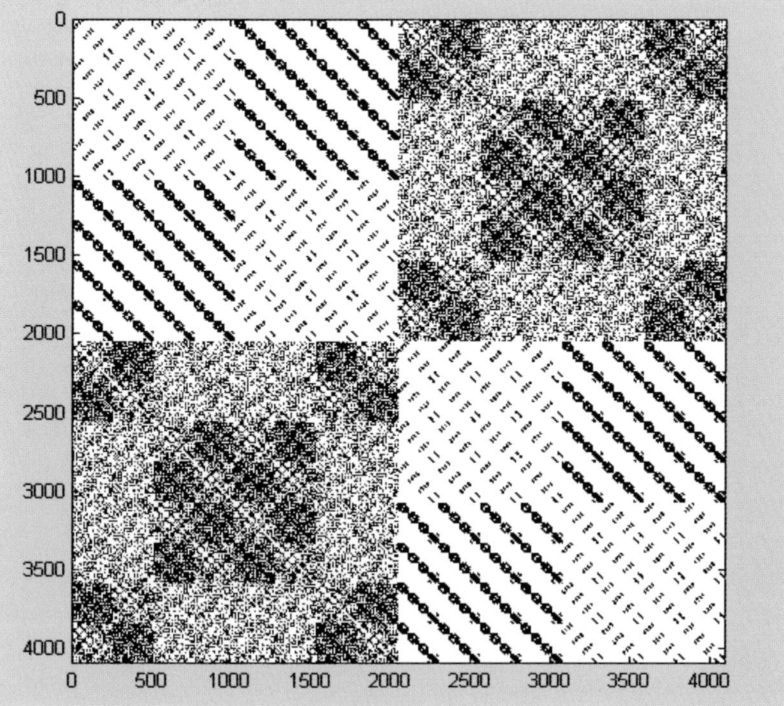

Fig. 6.6 QOC link connection between the $N = 4096$ seed matrices formed in step (a) for $P = N_t = 4$, rank 4 and weight 2

One of the efficient representations of sparse matrix is the graph model. This suggests that the code search/grouping problem in step (b) can be solved with the help of graph theory and graph-based algorithms. Specifically, the N matrices found in step (a) are modeled as a series of N nodes in a graph. For every pair of matrices that satisfy the QOC in (3.20), their nodes will be connected by a link. Hence a graph with nodes representing possible QO-STBC dispersion matrices, connected by links denoting conformance to the QOC between the connected nodes, will be formed. A simple example of such a graph is shown in Fig. 6.7, where the matrix $\mathbf{A}_1$ is assumed to satisfy QOC with matrices $\mathbf{A}_2$, $\mathbf{A}_4$ and $\mathbf{A}_5$; while the matrix $\mathbf{A}_3$ is assumed to satisfy QOC with matrices $\mathbf{A}_2$ and $\mathbf{A}_4$; and so on.

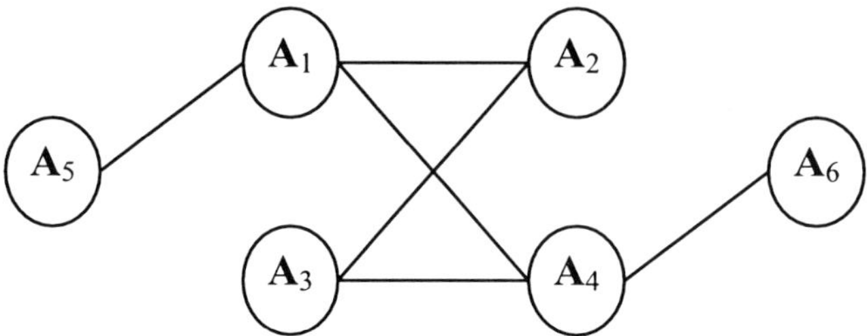

Fig. 6.7 Example of a graph

The graph example in Fig. 6.7 suggests that $\mathbf{A}_1$, $\mathbf{A}_2$, $\mathbf{A}_3$ and $\mathbf{A}_4$ can form a QO-STBC with dispersion matrices $\mathbf{A}_1$ and $\mathbf{A}_3$ in a group, and dispersion matrices $\mathbf{A}_2$ and $\mathbf{A}_4$ in another group. These two groups of dispersion matrices are separable from each other, because $\mathbf{A}_1$ and $\mathbf{A}_3$ establish the QOC link with $\mathbf{A}_2$ and $\mathbf{A}_4$, and vice versa. In a QO-STBC, since the dispersion matrices in *any* group must satisfy the QOC with all dispersion matrices in *all* other groups, in the graph model a QO-STBC will form a *fully connected* graph with nodes representing its dispersion matrices and links connecting every dispersion matrices of different groups. By exploiting this property, if a matrix node in this graph is randomly picked as the starting point to perform a "spanning tree algorithm" [86], one will be able to identify the quasi-orthogonal grouping of the N matrices obtained from step (a) effectively.

Depth First Search (DFS) [86] is an algorithm in graph theory that provides a systematic way to visit all the nodes in a graph from any starting node and form a spanning tree. The pseudo codes of DFS algorithm are given in Fig. 6.8. In order to solve the dispersion matrix grouping problem in step (b), the DFS algorithm is extended to a Modified DFS (MDFS) algorithm. The pseudo codes of the proposed MDFS algorithm are given in Fig. 6.9. Essential differences between the DFS and MDFS algorithms are listed below:

- In MDFS, every node can be visited more than once. In DFS, every node can only be visited once.
- In MDFS, there is an assignment of group to the nodes visited. There is no such assignment in DFS.

- In MDFS, there is an additional requirement that every visited node must be connected with its ancestors of different groups. There is no such requirement in DFS.

Input: Graph, number of nodes in graph (N)
Output: a spanning tree T with every node visited only once
Begin
    Pick a node n, $n \leq N$ as starting point;
    Assign node n as the root of tree T, i.e. $T = T \cup \{n\}$;
    DFS(n, T);
End

Procedure DFS(n, T)
Begin
    for ($v \in \{$neighbor of $n\}$)
        if ($v \notin T$)   % i.e. v not in the tree T
            Add node v to the tree T with n as parent, i.e. $T = T \cup \{v\}$;
            DFS(v, T);
        end
    end
End

Fig. 6.8 Depth First Search (DFS) Algorithm

Input: Graph, number of nodes in graph (N),
      number of groups required (G)
Output: a tree T with every branch as a valid solution
Begin
    Pick a node n, $n \leq N$ as starting point;
    Assign node n to Group 1, i.e. $g(n) = 1$;
    Assign node n as the root of tree T, i.e. $T = T \cup \{n\}$;
    MDFS(n, T, g);
End

Procedure MDFS(n, T, g)
Begin
    for ($v \in \{$neighbor of $n\}$)
        if ($v \notin \{$ancestor of $n\}$)
            Assume that node v is in Group p
           where $p = g(n)+1$ if $g(n)+1 \leq G$,
              $= 1$ if $g(n)+1 > G$;

           if (v possesses a link with all ancestors of n with
                different groups, i.e. g(ancestor of n) $\neq p$)
              Assign node v to Group p, i.e. $g(v) = p$;
              Add node v to the tree T with n as parent,
                i.e. $T = T \cup \{v\}$;
              MDFS(v, T, g);
           end
        end
    end
END

Fig. 6.9 Modified Depth First Search (MDFS) Algorithm

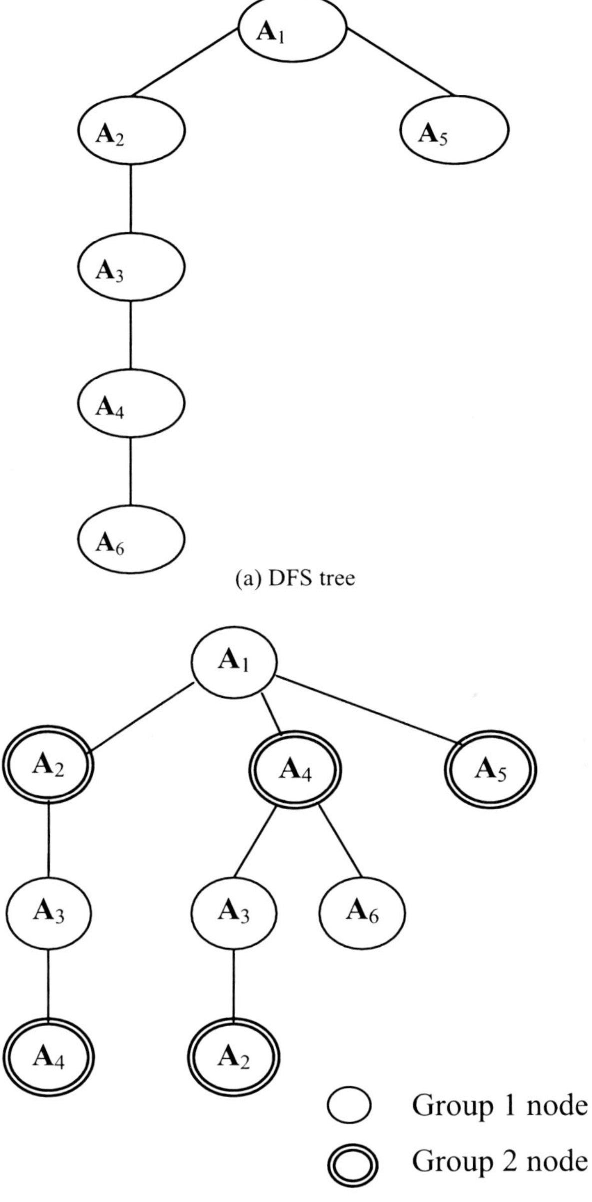

(a) DFS tree

(b) MDFS tree

Fig. 6.10 Trees generated by DFS and MDFS algorithms

The trees constructed by DFS and MDFS with $G = 2$ from the graph example in Fig. 6.7 are shown in Fig. 6.10(a) and (b) respectively. Every branch of the tree constructed by MDFS constitutes a possible solution for the dispersion matrices of a QO-STBC. For example in Fig. 6.10(b), $\mathbf{A}_1$-$\mathbf{A}_2$-$\mathbf{A}_3$-$\mathbf{A}_4$ and $\mathbf{A}_1$-$\mathbf{A}_4$-$\mathbf{A}_6$ and $\mathbf{A}_1$-$\mathbf{A}_5$ are possible solutions, but $\mathbf{A}_1$-$\mathbf{A}_4$-$\mathbf{A}_3$-$\mathbf{A}_2$ is not as it is merely a permutation of the first branch.

On the other hand, the $\mathbf{A}_1$-$\mathbf{A}_2$-$\mathbf{A}_3$-$\mathbf{A}_4$-$\mathbf{A}_6$ branch in the DFS tree in Fig. 6.10(a) is not a valid QO-STBC solution because although $\mathbf{A}_6$ has a QOC link with $\mathbf{A}_4$ (which is in group 2), it does not have a QOC link with its ancestor node $\mathbf{A}_2$ (which is also in group 2), hence $\mathbf{A}_6$ cannot be added as group 1 node and cannot form a QO-STBC together with $\mathbf{A}_2$ and $\mathbf{A}_4$. This explains why the basic DFS algorithm cannot be used for solving the code search problem described in step (b).

Back to Fig. 6.10(b), since a QO-STBC with as high code rate as possible is desired, the $\mathbf{A}_1$-$\mathbf{A}_2$-$\mathbf{A}_3$-$\mathbf{A}_4$ branch is picked as it gives a QO-STBC with the largest number of dispersion matrices. The resultant QO-STBC has a group of dispersion matrices consisting of $\mathbf{A}_1$ & $\mathbf{A}_3$, and another group of dispersion matrices consisting of $\mathbf{A}_2$ & $\mathbf{A}_4$. So a total of $M = 4$ dispersion matrices, divided into two orthogonal groups, are found, which agrees with the observation made from Fig. 6.7. Hence the MDFS algorithm can be used for solving the code search problem discussed in step (b) efficiently.

6.2.4   *Code search results*

Using the MDFS algorithm with $G = 2$ on the set of $N = 4096$ matrices with rank 4 and weight 2 described earlier, a few set of solutions can be found with $M = 16$ matrices grouped into two orthogonal groups. Each of these solution sets forms an equivalent channel matrix with a rank of $\mathcal{R} = 10$, resulting in a QO-STBC for four transmit antennas with code length $P = 4$, code rate $= \mathcal{R}/2T = 5/4$ and maximal symbol-wise diversity. One of such solution sets is given in Table 6.3 and will be used to discuss the relationship between rank and code rate of the equivalent channel matrix.

From the matrices $\mathbf{A}_1$ to $\mathbf{A}_{16}$ shown in Table 6.3, one can easily verify that the matrices $\mathbf{A}_1$ to $\mathbf{A}_8$ satisfy the QOC with the matrices $\mathbf{A}_9$ to $\mathbf{A}_{16}$. In other words, they may form a QO-STBC with two quasi-orthogonal groups. However, although $\mathbf{A}_1$ to $\mathbf{A}_{16}$ are 16 different matrices, the equivalent channel matrix formed by them has a rank of only 10, instead of 16. In other words, only 10 out of these 16 matrices ($\mathbf{A}_1$ to $\mathbf{A}_5$ and $\mathbf{A}_9$ to $\mathbf{A}_{13}$) are linearly independent and can be used as dispersion matrices to carry distinct information symbols, while the other 6 matrices ($\mathbf{A}_6$ to $\mathbf{A}_8$ and $\mathbf{A}_{14}$ to $\mathbf{A}_{16}$) are linearly dependent on the earlier 10 matrices and hence cannot be used to carry any new information symbol. Since every dispersion matrix carry one real information symbol and the matrices in Table 6.3 have length 4, the resultant QO-STBC has a code rate of 10/2/4 = 5/4.

The codeword of the rate-5/4 QO-STBC (based on dispersion matrices $\mathbf{A}_1$ to $\mathbf{A}_5$ and $\mathbf{A}_9$ to $\mathbf{A}_{13}$) is shown below:

$$\mathbf{C}_{rate-5/4} = \begin{bmatrix} a-d_6+d_7+jd_8+jd_9+jd_{10} & a+d_6-d_7+jd_8+jd_9-jd_{10} \\ a+d_6-d_7+jd_8-jd_9+jd_{10} & -a+d_6-d_7-jd_8+jd_9+jd_{10} \\ 0 & 0 \\ 0 & 0 \end{bmatrix} \cdots$$

$$\cdots \begin{bmatrix} 0 & 0 \\ 0 & 0 \\ d_1-d_2+jd_3+jd_4-jd_5-b & d_1-d_2+jd_3-jd_4+jd_5+b \\ d_1-d_2-jd_3+jd_4+jd_5+b & -d_1+d_2+jd_3+jd_4+jd_5+b \end{bmatrix},$$

$$(6.3)$$

where d_1 to d_{10} are ten real-valued code symbols, and $a = d_1+d_2+d_3+d_4+d_5$, $b = d_6+d_7+d_8+d_9+d_{10}$. Since 10 real symbols (equivalent to 5 complex symbols) are transmitted over 4 time units, rate 5/4 is achieved. The ML decoding of this code can be performed by two parallel decoders: one jointly detecting only d_1 to d_5, the other d_6 to d_{10}.

In addition to the code shown above, three other rate-5/4 QO-STBCs have been found using the aforementioned code search technique. Their dispersion matrices are listed in Appendix B.

Table 6.3 Matrices found by MDFS using $G = 2$, $P = N_t = 4$, rank $= 4$, weight $= 2$

<u>*Group 1*</u>

$$\mathbf{A}_1 = \begin{bmatrix} 1 & 1 & 0 & 0 \\ 1 & -1 & 0 & 0 \\ 0 & 0 & 1 & 1 \\ 0 & 0 & 1 & -1 \end{bmatrix}, \quad \mathbf{A}_2 = \begin{bmatrix} 1 & 1 & 0 & 0 \\ 1 & -1 & 0 & 0 \\ 0 & 0 & -1 & -1 \\ 0 & 0 & -1 & 1 \end{bmatrix},$$

$$\mathbf{A}_3 = \begin{bmatrix} 1 & 1 & 0 & 0 \\ 1 & -1 & 0 & 0 \\ 0 & 0 & j & j \\ 0 & 0 & -j & j \end{bmatrix}, \quad \mathbf{A}_4 = \begin{bmatrix} 1 & 1 & 0 & 0 \\ 1 & -1 & 0 & 0 \\ 0 & 0 & j & -j \\ 0 & 0 & j & j \end{bmatrix},$$

$$\mathbf{A}_5 = \begin{bmatrix} 1 & 1 & 0 & 0 \\ 1 & -1 & 0 & 0 \\ 0 & 0 & -j & j \\ 0 & 0 & j & j \end{bmatrix}, \quad \mathbf{A}_6 = \begin{bmatrix} 1 & 1 & 0 & 0 \\ 1 & -1 & 0 & 0 \\ 0 & 0 & -j & -j \\ 0 & 0 & j & -j \end{bmatrix},$$

$$\mathbf{A}_7 = \begin{bmatrix} 1 & 1 & 0 & 0 \\ 1 & -1 & 0 & 0 \\ 0 & 0 & -j & j \\ 0 & 0 & -j & -j \end{bmatrix}, \quad \mathbf{A}_8 = \begin{bmatrix} 1 & 1 & 0 & 0 \\ 1 & -1 & 0 & 0 \\ 0 & 0 & j & -j \\ 0 & 0 & -j & -j \end{bmatrix}.$$

Table 6.3 Matrices found by MDFS using $G = 2$, $P = Nt = 4$, rank = 4, weight = 2 (cont)

$$\underline{Group\ 2}$$

$$\mathbf{A}_9 = \begin{bmatrix} -1 & 1 & 0 & 0 \\ 1 & 1 & 0 & 0 \\ 0 & 0 & -1 & 1 \\ 0 & 0 & 1 & 1 \end{bmatrix}, \quad \mathbf{A}_{10} = \begin{bmatrix} 1 & -1 & 0 & 0 \\ -1 & -1 & 0 & 0 \\ 0 & 0 & -1 & 1 \\ 0 & 0 & 1 & 1 \end{bmatrix},$$

$$\mathbf{A}_{11} = \begin{bmatrix} j & j & 0 & 0 \\ j & -j & 0 & 0 \\ 0 & 0 & -1 & 1 \\ 0 & 0 & 1 & 1 \end{bmatrix}, \quad \mathbf{A}_{12} = \begin{bmatrix} j & j & 0 & 0 \\ -j & j & 0 & 0 \\ 0 & 0 & -1 & 1 \\ 0 & 0 & 1 & 1 \end{bmatrix},$$

$$\mathbf{A}_{13} = \begin{bmatrix} j & -j & 0 & 0 \\ j & j & 0 & 0 \\ 0 & 0 & -1 & 1 \\ 0 & 0 & 1 & 1 \end{bmatrix}, \quad \mathbf{A}_{14} = \begin{bmatrix} j & j & 0 & 0 \\ j & -j & 0 & 0 \\ 0 & 0 & 1 & -1 \\ 0 & 0 & -1 & -1 \end{bmatrix},$$

$$\mathbf{A}_{15} = \begin{bmatrix} j & j & 0 & 0 \\ -j & j & 0 & 0 \\ 0 & 0 & 1 & -1 \\ 0 & 0 & -1 & -1 \end{bmatrix}, \quad \mathbf{A}_{16} = \begin{bmatrix} j & -j & 0 & 0 \\ j & j & 0 & 0 \\ 0 & 0 & 1 & -1 \\ 0 & 0 & -1 & -1 \end{bmatrix}.$$

Table 6.4 Code search results found using proposed MDFS algorithm

Parameters			
Matrix Rank (symbol-wise diversity)	Dispersion Matrix Weight	Number of Group	Max. Code Rate
4	1	4	1
4	1	2	1
4	2	2	5/4
2	2	8	1
2	2	2	4

Table 6.4 summarizes the findings of the aforementioned code search using various code parameters. One can see that rate-5/4 QO-STBCs exist with symbol-wise diversity level = 4 and group = 2, while rate-4 QO-STBC's exist with symbol-wise diversity level = 2 and group = 2. Other interesting observations include:

- All QO-STBCs with code rate greater than one (shaded rows in Table 6.4) have dispersion matrices separated into 2 groups and weights greater than 1.
- To achieve a higher code rate from 5/4 to 4, the rank of the dispersion matrices (hence the symbol-wise diversity level) is reduced from 4 to 2, i.e. full transmit diversity can no longer be achieved.
- The rate-4 diversity-2 length-4 QO-STBC that has been found turns out to be equivalent to the rate-4 diversity-2 length-2 non-orthogonal STBCs [13,87], and both of them have the same decoding complexity.

One of the solution sets found for the rate-4 QO-STBC is given in Table 6.5. The codeword is shown in (6.4).

$$\mathbf{C}_{rate-4} = \begin{bmatrix} c_1 + c_3 - c_5 + c_7 - c_9 - c_{11} + c_{13} + c_{15} & c_1 + c_3 + c_5 - c_7 + c_9 - c_{11} - c_{13} + c_{15} \\ c_1 + c_3 - c_5 + c_7 + c_9 + c_{11} - c_{13} - c_{15} & c_1 + c_3 + c_5 - c_7 - c_9 + c_{11} + c_{13} - c_{15} \\ c_2 - c_4 + c_6 + c_8 + c_{10} + c_{12} + c_{14} + c_{16} & c_2 - c_4 - c_6 - c_8 - c_{10} + c_{12} - c_{14} + c_{16} \\ c_2 - c_4 + c_6 + c_8 - c_{10} - c_{12} - c_{14} - c_{16} & c_2 - c_4 - c_6 - c_8 + c_{10} - c_{12} + c_{14} - c_{16} \end{bmatrix} \cdots$$

$$\cdots \begin{bmatrix} c_2 + c_4 - c_6 + c_8 + c_{10} - c_{12} + c_{14} + c_{16} & c_2 + c_4 + c_6 - c_8 - c_{10} - c_{12} - c_{14} + c_{16} \\ c_2 + c_4 - c_6 + c_8 - c_{10} + c_{12} - c_{14} - c_{16} & c_2 + c_4 + c_6 - c_8 + c_{10} + c_{12} + c_{14} - c_{16} \\ c_1 - c_3 - c_5 - c_7 - c_9 - c_{11} - c_{13} - c_{15} & c_1 - c_3 + c_5 + c_7 + c_9 - c_{11} + c_{13} - c_{15} \\ c_1 - c_3 - c_5 - c_7 + c_9 + c_{11} + c_{13} + c_{15} & c_1 - c_3 + c_5 + c_7 - c_9 + c_{11} - c_{13} + c_{15} \end{bmatrix}.$$

$$(6.4)$$

Table 6.5 Dispersion Matrices of rate-4 QO-STBC found by MDFS using $G = 2$, $P = N_t = 4$, rank = 2, weight = 2

Group 1

$$\mathbf{A}_1 = \begin{bmatrix} 1 & 1 & 0 & 0 \\ 1 & 1 & 0 & 0 \\ 0 & 0 & 1 & 1 \\ 0 & 0 & 1 & 1 \end{bmatrix}, \quad \mathbf{A}_2 = \begin{bmatrix} j & j & 0 & 0 \\ j & j & 0 & 0 \\ 0 & 0 & j & j \\ 0 & 0 & j & j \end{bmatrix}, \quad \mathbf{A}_3 = \begin{bmatrix} 0 & 0 & 1 & 1 \\ 0 & 0 & 1 & 1 \\ 1 & 1 & 0 & 0 \\ 1 & 1 & 0 & 0 \end{bmatrix},$$

$$\mathbf{A}_4 = \begin{bmatrix} 0 & 0 & j & j \\ 0 & 0 & j & j \\ j & j & 0 & 0 \\ j & j & 0 & 0 \end{bmatrix}, \quad \mathbf{A}_5 = \begin{bmatrix} 1 & 1 & 0 & 0 \\ 1 & 1 & 0 & 0 \\ 0 & 0 & -1 & -1 \\ 0 & 0 & -1 & -1 \end{bmatrix}, \quad \mathbf{A}_6 = \begin{bmatrix} j & j & 0 & 0 \\ j & j & 0 & 0 \\ 0 & 0 & -j & -j \\ 0 & 0 & -j & -j \end{bmatrix},$$

$$\mathbf{A}_7 = \begin{bmatrix} 0 & 0 & 1 & 1 \\ 0 & 0 & 1 & 1 \\ -1 & -1 & 0 & 0 \\ -1 & -1 & 0 & 0 \end{bmatrix}, \quad \mathbf{A}_8 = \begin{bmatrix} 0 & 0 & j & j \\ 0 & 0 & j & j \\ -j & -j & 0 & 0 \\ -j & -j & 0 & 0 \end{bmatrix}, \quad \mathbf{A}_9 = \begin{bmatrix} -1 & 1 & 0 & 0 \\ -1 & 1 & 0 & 0 \\ 0 & 0 & -1 & 1 \\ 0 & 0 & -1 & 1 \end{bmatrix},$$

$$\mathbf{A}_{10} = \begin{bmatrix} -j & j & 0 & 0 \\ -j & j & 0 & 0 \\ 0 & 0 & -j & j \\ 0 & 0 & -j & j \end{bmatrix}, \quad \mathbf{A}_{11} = \begin{bmatrix} 0 & 0 & -1 & 1 \\ 0 & 0 & -1 & 1 \\ 1 & -1 & 0 & 0 \\ 1 & -1 & 0 & 0 \end{bmatrix}, \quad \mathbf{A}_{12} = \begin{bmatrix} 0 & 0 & -j & j \\ 0 & 0 & -j & j \\ j & -j & 0 & 0 \\ j & -j & 0 & 0 \end{bmatrix},$$

$$\mathbf{A}_{13} = \begin{bmatrix} 1 & -1 & 0 & 0 \\ 1 & -1 & 0 & 0 \\ 0 & 0 & -1 & 1 \\ 0 & 0 & -1 & 1 \end{bmatrix}, \quad \mathbf{A}_{14} = \begin{bmatrix} j & -j & 0 & 0 \\ j & -j & 0 & 0 \\ 0 & 0 & -j & j \\ 0 & 0 & -j & j \end{bmatrix}, \quad \mathbf{A}_{15} = \begin{bmatrix} 0 & 0 & 1 & -1 \\ 0 & 0 & 1 & -1 \\ 1 & -1 & 0 & 0 \\ 1 & -1 & 0 & 0 \end{bmatrix},$$

$$\mathbf{A}_{16} = \begin{bmatrix} 0 & 0 & j & -j \\ 0 & 0 & j & -j \\ j & -j & 0 & 0 \\ j & -j & 0 & 0 \end{bmatrix}.$$

Table 6.5 Dispersion Matrices of rate-4 QO-STBC found by MDFS using $G = 2$, $P = Nt = 4$, rank = 2, weight = 2 (cont)

Group 2

$$\mathbf{A}_{17} = \begin{bmatrix} -1 & 1 & 0 & 0 \\ 1 & -1 & 0 & 0 \\ 0 & 0 & -1 & 1 \\ 0 & 0 & 1 & -1 \end{bmatrix}, \quad \mathbf{A}_{18} = \begin{bmatrix} -j & j & 0 & 0 \\ j & -j & 0 & 0 \\ 0 & 0 & -j & j \\ 0 & 0 & j & -j \end{bmatrix}, \quad \mathbf{A}_{19} = \begin{bmatrix} 0 & 0 & 1 & -1 \\ 0 & 0 & -1 & 1 \\ -1 & 1 & 0 & 0 \\ 1 & -1 & 0 & 0 \end{bmatrix},$$

$$\mathbf{A}_{20} = \begin{bmatrix} 0 & 0 & j & -j \\ 0 & 0 & -j & j \\ -j & j & 0 & 0 \\ j & -j & 0 & 0 \end{bmatrix}, \quad \mathbf{A}_{21} = \begin{bmatrix} -1 & -1 & 0 & 0 \\ 1 & 1 & 0 & 0 \\ 0 & 0 & -1 & -1 \\ 0 & 0 & 1 & 1 \end{bmatrix}, \quad \mathbf{A}_{22} = \begin{bmatrix} -j & -j & 0 & 0 \\ j & j & 0 & 0 \\ 0 & 0 & -j & -j \\ 0 & 0 & j & j \end{bmatrix},$$

$$\mathbf{A}_{23} = \begin{bmatrix} 0 & 0 & -1 & -1 \\ 0 & 0 & 1 & 1 \\ 1 & 1 & 0 & 0 \\ -1 & -1 & 0 & 0 \end{bmatrix}, \quad \mathbf{A}_{24} = \begin{bmatrix} 0 & 0 & -j & -j \\ 0 & 0 & j & j \\ j & j & 0 & 0 \\ -j & -j & 0 & 0 \end{bmatrix}, \quad \mathbf{A}_{25} = \begin{bmatrix} 1 & -1 & 0 & 0 \\ -1 & 1 & 0 & 0 \\ 0 & 0 & -1 & 1 \\ 0 & 0 & 1 & -1 \end{bmatrix},$$

$$\mathbf{A}_{26} = \begin{bmatrix} j & -j & 0 & 0 \\ -j & j & 0 & 0 \\ 0 & 0 & -j & j \\ 0 & 0 & j & -j \end{bmatrix}, \quad \mathbf{A}_{27} = \begin{bmatrix} 0 & 0 & 1 & -1 \\ 0 & 0 & -1 & 1 \\ 1 & -1 & 0 & 0 \\ -1 & 1 & 0 & 0 \end{bmatrix}, \quad \mathbf{A}_{28} = \begin{bmatrix} 0 & 0 & j & -j \\ 0 & 0 & -j & j \\ j & -j & 0 & 0 \\ -j & j & 0 & 0 \end{bmatrix},$$

$$\mathbf{A}_{29} = \begin{bmatrix} 1 & 1 & 0 & 0 \\ -1 & -1 & 0 & 0 \\ 0 & 0 & -1 & -1 \\ 0 & 0 & 1 & 1 \end{bmatrix}, \quad \mathbf{A}_{30} = \begin{bmatrix} j & j & 0 & 0 \\ -j & -j & 0 & 0 \\ 0 & 0 & -j & -j \\ 0 & 0 & j & j \end{bmatrix}, \quad \mathbf{A}_{31} = \begin{bmatrix} 0 & 0 & 1 & 1 \\ 0 & 0 & -1 & -1 \\ 1 & 1 & 0 & 0 \\ -1 & -1 & 0 & 0 \end{bmatrix},$$

$$\mathbf{A}_{32} = \begin{bmatrix} 0 & 0 & j & j \\ 0 & 0 & -j & -j \\ j & j & 0 & 0 \\ -j & -j & 0 & 0 \end{bmatrix}.$$

In Fig. 6.11, the achievable capacity, as defined in (1.18), of rate-3/4 O-STBC, rate-1 QO-STBC and the rate-5/4 QO-STBC found in this section are compared for a four transmit and two receive antennas system (as shown in (1.17), in order to achieve a code rate greater than one, the minimum number of transmit and receive antenna must be also greater than one). It can be seen that the STBC with a lower code rate achieve a lower capacity, and the rate-5/4 QO-STBC that has been found achieves the highest capacity at high SNR region.

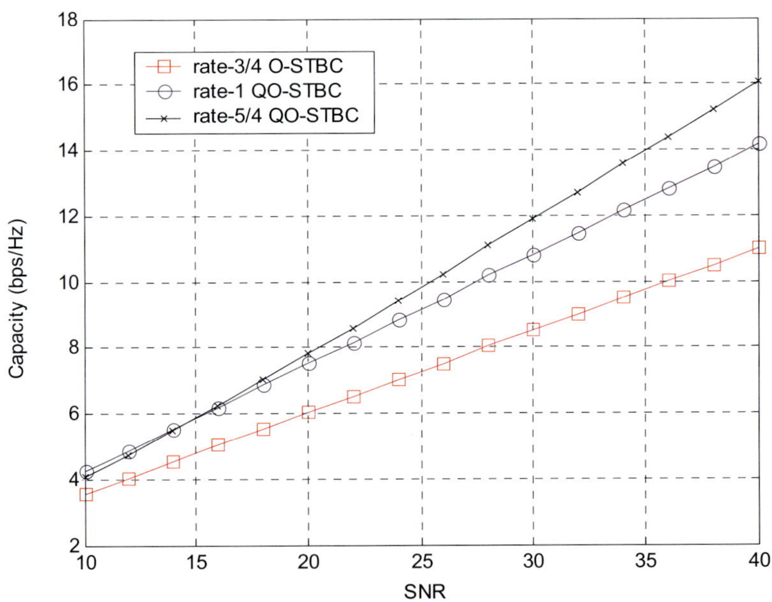

Fig. 6.11 Achievable capacity of STBC with four transmit and two receive antennas

6.2.5 *Section summary*

In this section, a graph-based approach incorporating the algebraic structure of the QO-STBC dispersion matrices presented in Chapter 3 is formulated to perform computer search for QO-STBC with code rate greater than one. Exploiting the sparse matrix approach, an efficient

Modified Depth First Search (MDFS) algorithm is developed to facilitate the code search. For 4 transmit antennas, a few QO-STBCs with rates 5/4 and 4 are found. A trade-off between the code rate and the symbol-wise transmit diversity level is observed in these high-rate QO-STBCs: the rate-5/4 codes have symbol-wise diversity of 4 while the rate-4 codes have reduced symbol-wise diversity of 2. All the high-rate QO-STBCs obtained are also found to have 2 quasi-orthogonal groups and the "weights" of these dispersion matrices are found to be two, i.e. there are two non-zero entries in every row of the dispersion matrices. To our knowledge, the rate-5/4 codes are the first known non-trivial QO-STBCs with code rate > 1.

6.3 Chapter Summary

In this chapter, a class of QO-STBC called 4Gp-QOSTBC that can always achieve rate 1 and full diversity for any number of transmit antennas is first presented. A good attribute of this code is that the transmitted symbols can be decoded in four parallel decoders, hence it can be considered to have the best tradeoff in decoding complexity and code rate when compared with other STBC codes reported in the literature. The construction rules of 4Gp-QOSTBC also provides an effective means to trade decoding complexity with any code rate less than one in well-defined steps. In the second part of this chapter, a graph-based search methodology for QO-STBC with code rate greater than 1 is presented, and its finding, the first QO-STBC with rate-5/4 for four transmit antennas, is reported. The search results suggest that in order to achieve a code rate greater than one, every code symbol should be transmitted on more than one transmit antenna at every time instant. Interestingly, the rate-complexity relationship of the rate > 1 codes is found to be consistent with the rate < 1 codes described in the first part of this chapter.

Chapter 7

Other Developments and Applications of QO-STBC

In this chapter, we will discuss some other developments of QO-STBC not included in the previous chapters. These include QO-STBC in a closed-loop system, concatenation of QO-STBC with forward error correction code, super space-time trellis code based on QO-STBC, and QO-STBC in frequency selective fading channel. The lower decoding complexity QO-STBC schemes (such as MDC-QOSTBC) discussed earlier in this monograph can be easily extended to these channels or systems. In the second part of this chapter, we will discuss the possible application of QO-STBC in future wireless communication standards, such as 3GPP LTE (3rd Generation Partnership Project — Long Term Evolution), a.k.a. the 4G mobile cellular standard. Since the standardization of 3GPP LTE is an on-going process during the time of writing this monograph, we can only review some proposals that include the application of QO-STBC. Finally, we shall give a conclusion for the monograph in the last section of this chapter.

7.1 Other Developments of QO-STBC

7.1.1 Closed-loop QO-STBC

In this monograph, we have so far focused on the feedforward and blind transmit diversity schemes of QO-STBC. However, when channel state information (CSI) is available to the transmitter, it can be exploited to improve the decoding performance or reduce the decoding complexity of QO-STBC. Such schemes are classified as feedback QO-STBC schemes. Different feedback schemes aimed at improving the decoding performance of QO-STBC have been reported in the past. Beam forming is considered in [88], antenna selection is used in [89], other approaches are reported in [90-94]. A feedback scheme to reduce decoding complexity of QO-STBC is proposed in [95].

In [96], an interesting retransmission scheme based on QO-STBC with 1-bit feedback (the receiver informs the transmitter whether the transmission is successful or failed) has been proposed as follows:

$$
\begin{aligned}
\text{1st transmission} \quad &: \mathbf{G}_0 = \begin{bmatrix} \mathbf{A} & \mathbf{B} \\ \mathbf{B} & \mathbf{A} \end{bmatrix} + \begin{bmatrix} \mathbf{C} & \mathbf{D} \\ -\mathbf{D} & -\mathbf{C} \end{bmatrix}; \\[2mm]
\text{2nd transmission} \quad &: \mathbf{G}_1 = \begin{bmatrix} \mathbf{A} & \mathbf{B} \\ \mathbf{B} & \mathbf{A} \end{bmatrix} - \begin{bmatrix} \mathbf{C} & \mathbf{D} \\ -\mathbf{D} & -\mathbf{C} \end{bmatrix}; \\[2mm]
\text{3rd transmission} \quad &: \mathbf{G}_2 = \begin{bmatrix} \mathbf{A} & -\mathbf{B} \\ -\mathbf{B} & \mathbf{A} \end{bmatrix} + \begin{bmatrix} \mathbf{C} & -\mathbf{D} \\ \mathbf{D} & -\mathbf{C} \end{bmatrix}; \\[2mm]
\text{4th transmission} \quad &: \mathbf{G}_3 = \begin{bmatrix} \mathbf{A} & -\mathbf{B} \\ -\mathbf{B} & \mathbf{A} \end{bmatrix} - \begin{bmatrix} \mathbf{C} & -\mathbf{D} \\ \mathbf{D} & -\mathbf{C} \end{bmatrix},
\end{aligned}
\tag{7.1}
$$

where $\mathbf{A}$, $\mathbf{B}$, $\mathbf{C}$, $\mathbf{D}$ are each an Alamouti STBC.

The 1st transmission is a rate-2 code called Double-ABBA (DABBA) $\mathbf{G}_0$, which is the sum of two ABBA codes (ABBA is the TBH code specified in (2.21) in this monograph). If reception error occurs, the transmitter carries out a 2nd transmission using $\mathbf{G}_1$, the receiver decodes the sum of $\mathbf{G}_0$ and $\mathbf{G}_1$. By combining $\mathbf{G}_0$ and $\mathbf{G}_1$, one obtains two independent rate-1 ABBA codes, which are more robust than the first received DABBA. If reception error occurs again, subsequent retransmissions will send $\mathbf{G}_2$ and $\mathbf{G}_3$ respectively. After combining $\mathbf{G}_0$ to $\mathbf{G}_3$, the resultant codes are four rate-1/2 O-STBC, which are very robust and simple to decode. In short, for every retransmission, a different STBC is transmitted so as to be combined at the receiver with previously received STBC's to achieve incremental detection robustness. Such scheme is also called Hybrid Automatic Repeat reQuest (HARQ).

7.1.2 *Concatenation of QO-STBC with error correction code*

In practical system, the data stream will be encoded with forward error correction (FEC) code. The combination of FEC with QO-STBC has been investigated in [92,97,98] by exploiting the quasi-orthogonality properties of QO-STBC. The main idea is to combine the error correction capability of FEC with the transmit diversity provided by the QO-STBC.

7.1.3 *Super space-time trellis code based on QO-STBC*

Super Orthogonal STTC (Space-Time Trellis Code) and Super Quasi-Orthogonal STTC have been proposed in [99,100]. These STTCs use O-STBC and QO-STBC as the codewords of a STTC, in order to achieve higher coding gain than that achievable by a STBC and lower decoding complexity than that achievable by a traditional STTC. The extension of this idea to MDC-QOSTBC, herein called Super MDC-QO STTC, promises to further lower the decoding complexity than Super Quasi-Orthogonal STTC [101].

Table 7.1 compares the decoding complexity of two Super STTC schemes, namely Super Quasi-Orthogonal STTC from [100] and Super MDC-QO STTC from [101]. The decoding of these Super STTC schemes can generally be performed in two steps: The first step is to calculate the branch metrics and find a survivor branch for each state transition. Then Viterbi algorithm is applied on these survivor branches to find the survivor path. Since there are many multiplication operations in the first step, while there are only simple operations such as comparison and addition in the second step, the nett decoding complexity is mainly determined by the complexity of the first step [101]. As shown in the middle column of Table 7.1, Super MDC-QO STTC has a lower decoding complexity than Super Quasi-Orthogonal STTC, and this decoding complexity advantage increases with the constellation size.

Table 7.1 Decoding complexity comparison of different Super STTC schemes

Super STTC based on	Likelihood function calculations per symbol	State transitions per symbol interval
MDC-QOSTBC [101]		
2 states (4QAM)	8	2
4 states (4QAM)	8	4
QO-STBC [100]		
2 states (4QAM)	16	1
4 states (4QAM)	16	2

7.1.4 QO-STBC in frequency selective fading channel

The QO-STBC designs in this monograph have been aimed for frequency flat fading channel. QO-STBC has been extended to frequency-selective fading channel in [98,102-106] to form the so-called Space-Frequency Code or Space-Time-Frequency Code. Similar extension of MDC-QOSTBC to the frequency selective fading channels will be straightforward.

7.2 QO-STBC in Communication Standards

With the successful deployment of 3G cellular systems around the world, the standardization of 4G cellular systems, or 3GPP LTE, has been carried out since 2005 [107]. The main objective is "to develop a framework for the evolution of the 3GPP radio-access technology towards a high-data-rate, low-latency and packet-optimized radio-access technology". In the radio-interface physical layer of 3GPP LTE, "advanced multi-antenna technologies" is specified as an important area that needs detailed investigation.

In [108] and [109], Motorola and Samsung show that double–symbol–decodable QO-STBC has the best performance used with four transmit and one receive antennas. However, both also express concerns over the decoding complexity of QO-STBC, as double–symbol joint detection is still considered highly complex as compared with CSD (cyclic shift diversity) or CDD (cyclic delay diversity). Hence QO-STBC with a lower decoding complexity, such as the QO-STBC with GCLT or MDC-QOSTBC described in this monograph, are expected to be good candidates for four-antenna transmit diversity.

In [110], Huawei gave a further comparison between QO-STBC with CSD in coded systems. They show that QO-STBC provides a better performance than CSD at low FEC rate, while the reverse is true at high FEC rate.

In [111] and [112], Nortel Network propose a HARQ (hybrid ARQ) setting based on the three STBCs shown in (7.2): a rate-1 QO-STBC (which is the TBH code shown in (2.21)), a rate-2 STBC called DSTTD, and a rate-4 SM (spatial multiplexing) code. It can be seen that the SM code is exactly the first row of the DSTTD code, while the DSTTD code is exactly the first two rows of the QO-STBC. This makes them suitable for use in HARQ mode.

The HARQ mode works as follow: we first transmit four data streams using the rate-4 SM code. If error occurs, we can transmit the second row of DSTTD, and the receiver receives an equivalent rate-2 DSTTD code. By doing so, the transmit diversity level is increased from one to two (while the effective code rate is reduced from 4 to 2). If error still occurs, we next transmit the last two rows of QO-STBC, and the receiver received an equivalent rate-1 QO-STBC. By doing so, the transmit diversity is further increased to four (while the effective code rate is reduced to 1).

The HARQ principle is as follow: If the channel condition is good, a rate-4 code with transmit diversity one is sufficient and will achieve a high throughput. If the channel condition is bad, a rate-1 code with transmit diversity four will provide additional robustness against fading, though at a lower throughput due to retransmissions. Since the code rate of the HARQ system is adapted to the channel condition, its overall throughput is expected to be higher than a non-adaptive system.

rate-4:
$$SM = \begin{bmatrix} c_1 & c_2 & c_3 & c_4 \end{bmatrix}.$$

rate-2:
$$DSTTD = \begin{bmatrix} \mathbf{A} & \mathbf{B} \end{bmatrix} = \begin{bmatrix} c_1 & c_2 & c_3 & c_4 \\ -c_2^* & c_1^* & -c_4^* & c_3^* \end{bmatrix}. \tag{7.2}$$

rate-1:
$$QO - STBC = \begin{bmatrix} \mathbf{A} & \mathbf{B} \\ \mathbf{B} & \mathbf{A} \end{bmatrix} = \begin{bmatrix} c_1 & c_2 & c_3 & c_4 \\ -c_2^* & c_1^* & -c_4^* & c_3^* \\ c_3 & c_4 & c_1 & c_2 \\ -c_4^* & c_3^* & -c_2^* & c_1^* \end{bmatrix}.$$

In [113], LG proposed a new rate-2 code for two transmit antennas. It is called **XTD** and is shown in (7.3). Comparing (7.3) with MDC-QOSTBC in (4.17), which is reproduced in (7.4), some similarity can be observed: The boxed term in (7.4) is exactly the same as (7.3), after reindexing and negating some of the symbols.

$$\mathbf{XTD} = \begin{bmatrix} x_1^R + jx_4^I & x_2^R + jx_3^I \\ x_3^R + jx_2^I & x_4^R + jx_1^I \end{bmatrix}. \tag{7.3}$$

$$\mathbf{MDC\text{-}QOSTBC} = \begin{bmatrix} \boxed{c_1^R + jc_3^I} & c_2^R + jc_4^I & -c_3^R - jc_1^I & \boxed{-c_4^R - jc_2^I} \\ \boxed{-c_2^R + jc_4^I} & c_1^R - jc_3^I & -c_4^R + jc_2^I & \boxed{c_3^R - jc_1^I} \\ c_3^R + jc_1^I & c_4^R + jc_2^I & c_1^R + jc_3^I & c_2^R + jc_4^I \\ c_4^R - jc_2^I & -c_3^R + jc_1^I & -c_2^R + jc_4^I & c_1^R - jc_3^I \end{bmatrix}. \tag{7.4}$$

Through the above case studies and examples, we can see that QO-STBC is receiving attention as an important technology for the 3G LTE/4G systems. The high-rate low-complexity full-diversity codes presented in this monograph, such as MDC-QOSTBC, are clearly good, if not better, candidates for these future wireless communication systems. In addition, as shown earlier, in addition to four transmit antennas, MDC-QOSTBC can be applied with two transmit antennas as well, which include the **XTD** proposal from LG as a special case.

Chapter 8

Conclusions

In this monograph, we have focused on the design, analysis, construction and performance optimization of coherent and non-coherent transmit diversity schemes based on Quasi-Orthogonal Space-Time Block Code (QO-STBC).

We have derived the algebraic structure of QO-STBC to unify the structural requirements of all generic QO-STBC in Chapter 3. We have also derived the generalized whitening filter that permits the decoding of QO-STBC using standard decoding or equalization schemes. With the derived algebraic structure of QO-STBC, we have found that although the well known constellation rotation (CR) helps QO-STBC to attain full diversity, it actually increases its decoding complexity when regular square or rectangular quadrature amplitude modulation (QAM) constellations are used. Therefore we proposed a new constellation transformation technique called Group-Constrained Linear Transformation (GCLT) to optimise QO-STBC without any increase in decoding complexity. We have also derived the optimum GCLT parameters for QO-STBC with square- and rectangular-QAM constellations to achieve full diversity and maximum coding gain. The optimised QO-STBC with GCLT has only very slight drop in decoding performance compared to QO-STBC with CR.

As GCLT can only be applied to QO-STBC with square- and rectangular-QAM constellations, we have further proposed in Chapter 4 the MDC-QOSTBC (QO-STBC with minimum decoding complexity) which can be used with any constellation, and derived its algebraic structure requirements. MDC-QOSTBC is single-symbol decodable, i.e.

its ML decodings only need to jointly detect two real symbols, hence it has the lowest decoding complexity among all QO-STBC, and a decoding complexity higher only to O-STBC. We have proposed a new concept of Preferred AOD Pair to link MDC-QOSTBC to AOD, and to derive the theoretical achievable code rate of MDC-QOSTBC: 1 for four transmit antennas and 3/4 for eight transmit antennas. We have also designed systematic construction rules to construct MDC-QOSTBC from Preferred AOD Pair or O-STBC, and derived the optimum constellation rotation angle for these MDC-QOSTBC. For square- and rectangular-QAM, the optimum constellation rotation angle is found to be 13.3^0 for every symbol. The proposed MDC-QOSTBC is shown to achieve full diversity, higher code rate than O-STBC, and better power distribution as well as better flexibility to support different number of transmit antennas than the Coordinate Interleaved Orthogonal Design (CIOD) and Asymmetry CIOD (ACIOD) codes.

In an effort to develop new blind transmit diversity schemes with high code rate and low decoding complexity, in Chapter 5 we have designed new Differential Space-Time Modulation (DSTM) schemes based on double–symbol–decodable QO-STBC and single–symbol–decodable MDC-QOSTBC. We have proposed the use of joint modulation and specially designed constellation set such that a set of unitary/quasi-unitary matrices can be constructed from QO-STBC/MDC-QOSTBC. DSTM schemes based on these unitary/quasi-unitary matrices have very low decoding complexity because they can achieve ML decoding by jointly decoding only two complex symbols (double–symbol–decodable)/two real symbols (single–symbol–decodable) respectively. These are the lowest ML decoding complexity ever reported for DSTM schemes not based on O-STBC. Compared with DSTM scheme based on O-STBC, the proposed DSTMs schemes can achieve a higher code rate and a better decoding performance. Compared with the DSTM based on Sp(2), the proposed DSTMs have only a slight degradation in decoding performance but much lower decoding search space.

Finally, we have discussed full-diversity QO-STBC with less stringent requirements on decoding complexity but more emphasis on scalability and flexibility in Chapter 6. We have designed a class of QO-

STBC called 4Gp-QOSTBC that can always achieve full rate and full diversity for any number of transmit antennas, hence they can be applied in any system without the need for rate matching. 4Gp-QOSTBC achieves a lower decoding complexity than the existing full-rate high-diversity codes as it always decouples the transmitted symbols into four quasi-orthogonal groups. We have also formulated a graph-based code search methodology to look for QO-STBC with rate greater than one, and found the first QO-STBC with rate-5/4 for four transmit antennas.

The authors hope that the ideas collated in this monograph will help spur futher innovations in MIMO and space-time processing to make future wireless communications more reliable and exciting.

APPENDIX A

Properties of Kronecker Product

Let's assume two matrices $\mathbf{X}$ and $\mathbf{Y}$ as follows:

$$\mathbf{X} = \begin{bmatrix} x_{1,1} & x_{1,2} \\ x_{2,1} & x_{2,2} \end{bmatrix}, \qquad \mathbf{Y} = \begin{bmatrix} y_{1,1} & y_{1,2} \\ y_{2,1} & y_{2,2} \end{bmatrix}. \qquad (A.1)$$

The Kronecker product, $\otimes$, is defined as follows:

$$\begin{aligned}
\mathbf{X} \otimes \mathbf{Y} &= \begin{bmatrix} x_{1,1}\mathbf{Y} & x_{1,2}\mathbf{Y} \\ x_{2,1}\mathbf{Y} & x_{2,2}\mathbf{Y} \end{bmatrix} \\
&= \begin{bmatrix} x_{1,1}y_{1,1} & x_{1,1}y_{1,2} & x_{1,2}y_{1,1} & x_{1,2}y_{1,2} \\ x_{1,1}y_{2,1} & x_{1,1}y_{2,2} & x_{1,2}y_{2,1} & x_{1,2}y_{2,2} \\ x_{2,1}y_{1,1} & x_{2,1}y_{1,2} & x_{2,2}y_{1,1} & x_{2,2}y_{1,2} \\ x_{2,1}y_{2,1} & x_{2,1}y_{2,2} & x_{2,2}y_{2,1} & x_{2,2}y_{2,2} \end{bmatrix}.
\end{aligned} \qquad (A.2)$$

Some properties of the Kronecker product, $\otimes$, are listed below:

$$\begin{aligned}
&\text{(i)} \quad (\mathbf{X} \otimes \mathbf{Y})(\mathbf{W} \otimes \mathbf{Z}) = (\mathbf{XW}) \otimes (\mathbf{YZ}); \\
&\text{(ii)} \quad (\mathbf{X} \otimes \mathbf{Y})^{H} = (\mathbf{X}^{H}) \otimes (\mathbf{Y}^{H}); \\
&\text{(iii)} \quad (\mathbf{X} \otimes \mathbf{Y}) + (\mathbf{X} \otimes \mathbf{Y})^{H} = 0; \\
&\qquad\qquad \text{if} \quad \mathbf{X}=\mathbf{X}^{H} \ \& \ \mathbf{Y}=-\mathbf{Y}^{H}, \\
&\qquad\qquad \text{or if} \quad \mathbf{X}=-\mathbf{X}^{H} \ \& \ \mathbf{Y}=\mathbf{Y}^{H}, \\
&\text{(iv)} \quad (\mathbf{X} \otimes \mathbf{Y}) - (\mathbf{X} \otimes \mathbf{Y})^{H} = 0; \\
&\qquad\qquad \text{if} \quad \mathbf{X}=\mathbf{X}^{H} \ \& \ \mathbf{Y}=\mathbf{Y}^{H}, \\
&\qquad\qquad \text{or if} \quad \mathbf{X}=-\mathbf{X}^{H} \ \Rightarrow \ \mathbf{Y}=-\mathbf{Y}^{H}.
\end{aligned} \qquad (A.3)$$

Properties (A.3) (i) and (ii) can be found in [114], while (A.3) (iii) and (iv) can be easily verified.

APPENDIX B

Dispersion Matrices of Rate-5/4 QO-STBC

In this appendix, we list the dispersion matrices of three other rate-5/4 QO-STBC arising from the code search described in Section 6.2.

Code 1

Group 1

$$\mathbf{A}_1 = \begin{bmatrix} 1 & 1 & 0 & 0 \\ 1 & -1 & 0 & 0 \\ 0 & 0 & 1 & 1 \\ 0 & 0 & 1 & -1 \end{bmatrix},$$

$$\mathbf{A}_4 = \begin{bmatrix} 1 & 1 & 0 & 0 \\ 1 & -1 & 0 & 0 \\ 0 & 0 & -1 & -1 \\ 0 & 0 & -1 & 1 \end{bmatrix},$$

$$\mathbf{A}_3 = \begin{bmatrix} -j & j & 0 & 0 \\ -j & -j & 0 & 0 \\ 0 & 0 & -1 & -1 \\ 0 & 0 & -1 & 1 \end{bmatrix},$$

$$\mathbf{A}_4 = \begin{bmatrix} j & -j & 0 & 0 \\ -j & -j & 0 & 0 \\ 0 & 0 & -1 & -1 \\ 0 & 0 & -1 & 1 \end{bmatrix},$$

$$\mathbf{A}_5 = \begin{bmatrix} j & j & 0 & 0 \\ -j & j & 0 & 0 \\ 0 & 0 & -1 & -1 \\ 0 & 0 & -1 & 1 \end{bmatrix},$$

Group 2

$$\mathbf{A}_6 = \begin{bmatrix} 1 & -1 & 0 & 0 \\ -1 & -1 & 0 & 0 \\ 0 & 0 & -j & -j \\ 0 & 0 & -j & j \end{bmatrix},$$

$$\mathbf{A}_7 = \begin{bmatrix} 1 & -1 & 0 & 0 \\ -1 & -1 & 0 & 0 \\ 0 & 0 & 1 & -1 \\ 0 & 0 & -1 & -1 \end{bmatrix},$$

$$\mathbf{A}_8 = \begin{bmatrix} 1 & -1 & 0 & 0 \\ -1 & -1 & 0 & 0 \\ 0 & 0 & -j & -j \\ 0 & 0 & j & -j \end{bmatrix},$$

$$\mathbf{A}_9 = \begin{bmatrix} 1 & -1 & 0 & 0 \\ -1 & -1 & 0 & 0 \\ 0 & 0 & -1 & 1 \\ 0 & 0 & 1 & 1 \end{bmatrix},$$

$$\mathbf{A}_{10} = \begin{bmatrix} 1 & -1 & 0 & 0 \\ -1 & -1 & 0 & 0 \\ 0 & 0 & -j & j \\ 0 & 0 & -j & -j \end{bmatrix}.$$

Code 2

Group 1   Group 2

$$\mathbf{A}_1 = \begin{bmatrix} 1 & -j & 0 & 0 \\ 1 & j & 0 & 0 \\ 0 & 0 & 1 & -j \\ 0 & 0 & -j & 1 \end{bmatrix},$$

$$\mathbf{A}_6 = \begin{bmatrix} -1 & -j & 0 & 0 \\ 1 & -j & 0 & 0 \\ 0 & 0 & -1 & -j \\ 0 & 0 & -j & -1 \end{bmatrix},$$

$$\mathbf{A}_2 = \begin{bmatrix} 1 & -j & 0 & 0 \\ 1 & j & 0 & 0 \\ 0 & 0 & -1 & j \\ 0 & 0 & j & -1 \end{bmatrix},$$

$$\mathbf{A}_7 = \begin{bmatrix} 1 & j & 0 & 0 \\ -1 & j & 0 & 0 \\ 0 & 0 & -1 & -j \\ 0 & 0 & -j & -1 \end{bmatrix},$$

$$\mathbf{A}_3 = \begin{bmatrix} 1 & -j & 0 & 0 \\ 1 & j & 0 & 0 \\ 0 & 0 & j & -1 \\ 0 & 0 & -1 & j \end{bmatrix},$$

$$\mathbf{A}_8 = \begin{bmatrix} -j & 1 & 0 & 0 \\ -j & -1 & 0 & 0 \\ 0 & 0 & -1 & -j \\ 0 & 0 & -j & -1 \end{bmatrix},$$

$$\mathbf{A}_4 = \begin{bmatrix} 1 & -j & 0 & 0 \\ 1 & j & 0 & 0 \\ 0 & 0 & j & -1 \\ 0 & 0 & 1 & -j \end{bmatrix},$$

$$\mathbf{A}_9 = \begin{bmatrix} j & 1 & 0 & 0 \\ j & -1 & 0 & 0 \\ 0 & 0 & 1 & j \\ 0 & 0 & j & 1 \end{bmatrix},$$

$$\mathbf{A}_5 = \begin{bmatrix} 1 & -j & 0 & 0 \\ 1 & j & 0 & 0 \\ 0 & 0 & j & 1 \\ 0 & 0 & -1 & -j \end{bmatrix},$$

$$\mathbf{A}_{10} = \begin{bmatrix} j & 1 & 0 & 0 \\ -j & 1 & 0 & 0 \\ 0 & 0 & -1 & -j \\ 0 & 0 & -j & -1 \end{bmatrix}.$$

Code 3

Group 1

$$\mathbf{A}_1 = \begin{bmatrix} 1 & -j & 0 & 0 \\ 1 & j & 0 & 0 \\ 0 & 0 & 1 & -j \\ 0 & 0 & -j & 1 \end{bmatrix},$$

$$\mathbf{A}_2 = \begin{bmatrix} 1 & -j & 0 & 0 \\ 1 & j & 0 & 0 \\ 0 & 0 & -1 & j \\ 0 & 0 & j & -1 \end{bmatrix},$$

$$\mathbf{A}_3 = \begin{bmatrix} 1 & -j & 0 & 0 \\ 1 & j & 0 & 0 \\ 0 & 0 & -j & 1 \\ 0 & 0 & 1 & -j \end{bmatrix},$$

$$\mathbf{A}_4 = \begin{bmatrix} 1 & -j & 0 & 0 \\ 1 & j & 0 & 0 \\ 0 & 0 & -j & 1 \\ 0 & 0 & -1 & j \end{bmatrix},$$

$$\mathbf{A}_5 = \begin{bmatrix} 1 & -j & 0 & 0 \\ 1 & j & 0 & 0 \\ 0 & 0 & -j & -1 \\ 0 & 0 & 1 & j \end{bmatrix},$$

Group 2

$$\mathbf{A}_6 = \begin{bmatrix} 1 & j & 0 & 0 \\ -1 & j & 0 & 0 \\ 0 & 0 & 1 & j \\ 0 & 0 & j & 1 \end{bmatrix},$$

$$\mathbf{A}_7 = \begin{bmatrix} 1 & j & 0 & 0 \\ -1 & j & 0 & 0 \\ 0 & 0 & -1 & -j \\ 0 & 0 & -j & -1 \end{bmatrix},$$

$$\mathbf{A}_8 = \begin{bmatrix} j & -1 & 0 & 0 \\ j & 1 & 0 & 0 \\ 0 & 0 & -1 & -j \\ 0 & 0 & -j & -1 \end{bmatrix},$$

$$\mathbf{A}_9 = \begin{bmatrix} j & 1 & 0 & 0 \\ j & -1 & 0 & 0 \\ 0 & 0 & -1 & -j \\ 0 & 0 & -j & -1 \end{bmatrix},$$

$$\mathbf{A}_{10} = \begin{bmatrix} j & 1 & 0 & 0 \\ -j & 1 & 0 & 0 \\ 0 & 0 & 1 & j \\ 0 & 0 & j & 1 \end{bmatrix}.$$

BIBLIOGRAPHY

[1]   E. Telatar, "Capacity of multi-antenna Gaussian channels", *European Transactions on Telecommunications,* vol. 10, pp. 585–595, Nov./Dec. 1999.

[2]   G. J. Foschini and M. J. Gans, "On limits of wireless communications in a fading environment when using multiple antennas", *Wireless Personal Communications,* vol. 6, pp. 311–335, 1998.

[3]   R. T. Derryberry, S. D. Gray, D. M. Ionescu, G. Mandyam, and B. Raghothaman, "Transmit diversity in 3G CDMA systems", *IEEE Communications Magazine,* vol. 40, pp.68–75, Apr. 2002.

[4]   V. Tarokh, N. Seshadri, and A. R. Calderbank, "Space-time codes for high data rate wireless communication: performance criterion and code construction", *IEEE Trans. on Information Theory*, vol. 44, pp. 744–765, Mar. 1998.

[5]   S. M. Alamouti, "A simple transmit diversity technique for wireless communications", *IEEE Journal on Selected Areas in Communications,* vol.16, pp.1451–1458, Oct. 1998.

[6]   V. Tarokh, H. Jafarkhani, and A. R. Calderbank, "Space-time block codes from orthogonal designs", *IEEE Trans. on Information Theory,* vol. 45, pp. 1456–1467, Jul. 1999.

[7]   S. Sandhu and A. Paulraj, "Space-time block codes: a capacity perspective", *IEEE Communications Letters*, vol. 4, pp. 384–386, Dec. 2000.

[8]   H. Jafarkhani, "A quasi-orthogonal space-time block code", *IEEE Trans. on Communications,* vol. 49, pp. 1–4, Jan. 2001.

[9]   O. Tirkkonen, A. Boariu, and A. Hottinen, "Minimal non-orthogonality rate 1 space-time block code for 3+ tx antennas", *IEEE ISSSTA 2000,* vol. 2, pp. 429–432.

[10] C. B. Papadias and G. J. Foschini, "Capacity-approaching space-time codes for systems employing four transmitter antennas", *IEEE Trans. on Information Theory*, vol. 49, pp. 726–733, Mar. 2003.

[11] B. Hassibi and B. M. Hochwald, "High-rate codes that are linear in space and time", *IEEE Trans. on Information Theory*, vol. 48, pp. 1804–1824, Jul. 2002.

[12] G. G. Giannakis, Y. Hua, P. Stoica, and L. Tong, *Signal Processing Advances in Wireless and Mobile Communications*, vol. 2, Prentice Hall PTR, 2001.

[13] A. Hottinen, O. Tirkkonen, and R. Wichman, *Multi-antenna Transceiver Techniques for 3G and Beyond*, John Wiley & Sons, 2003.

[14] J. C. Guey, M. P. Fitz, M. R. Bell, and W. Y. Kuo, "Signal design for transmitter diversity wireless communication systems over Rayleigh fading channels", *IEEE Trans. on Communications*, vol.47, pp. 527–537, Apr. 1999.

[15] J. Yuan, B. Vucetic, B. Xu, and Z. Chen, "Design of space-time codes and its performance in CDMA systems", *IEEE VTC-Spring 2001*, pp. 1292–1296.

[16] W. Su and X. Xia, "Signal constellations for quasi-orthogonal space-time block codes with full diversity", *IEEE Trans. Information Theory*, vol. 50, pp. 2331–2347, Oct 2004.

[17] G. Ganesan and P. Stoica, "Space-time block codes: a maximum SNR approach", *IEEE Trans. on Information Theory*, vol. 47, pp.1650–1656, May 2001.

[18] S. Sandhu and A. Paulraj, "Union bound on error probability of linear space-time block codes", *IEEE ICASSP 2001*, vol. 4, pp. 2473–2476.

[19] O. Tirkkonen, A. Hottinen, "Square-matrix embeddable space-time block codes for complex signal constellations", *IEEE Trans. on Information Theory*, vol. 48, pp.384–395, Feb. 2002.

[20] A. V. Geramita, J. M. Geramita, and J. S. Wallis, "Orthogonal designs", *Linear and Multilinear Algebra*, vol. 3, pp. 281–306, 1976.

[21] A. V. Geramita and J. Seberry, Orthogonal designs, quadratic forms and Hadamard matrices, Marcel Dekker, 1979.

[22] W. Wolfe, "Amicable orthogonal designs—existence", *Canadian Journal of Mathematics*, vol. XXVIII, pp. 1006–1020, 1976.

[23] C. Yuen, Y. L. Guan, and T. T. Tjhung, "Orthogonal space-time block code from amicable complex orthogonal design", *IEEE ICASSP 2004*, pp. 469–472.

[24] H. Wang, X. Xia, "Upper bounds of rates of complex orthogonal space-time block codes", *IEEE Trans. on Information Theory*, vol. 49, pp. 2788–2796, Oct. 2003.

[25] X. Liang and X. Xia, "On the nonexistence of rate-one generalized complex orthogonal designs", *IEEE Trans. on Information Theory*, vol. 49, pp. 2984–2988, Nov. 2003.

[26] W. Su, X. Xia, "Two generalized complex orthogonal space-time block codes of rates 7/11 and 3/5 for 5 and 6 transmit antennas", *IEEE Trans. on Information Theory*, vol. 49, pp. 313–316, Jan 2003.

[27] X. Liang, "Orthogonal designs with maximal rates", *IEEE Trans. on Information Theory*, vol. 49, pp. 2468–2503, Oct. 2003.

[28] K. Lu, S. Fu, and X. Xia, "Close forms designs of complex orthogonal space-time block codes of rates (k+1)/(2k) for 2k-1 or 2k transmit antennas", *IEEE Trans. on Information Theory*, vol. 51, pp. 4340–4347, Dec 2005.

[29] G. J. Foschini and C. B. Papadias, "Open-loop diversity technique for systems employing 4 transmitter antennas", *US Patent Application Publication*, US 2002/0118770.

[30] G. J. Foschini and C. B. Papadias, "Open-loop diversity technique for systems employing 4 transmitter antennas", *European Patent*, EP1223702.

[31] C. Yuen, Y. L. Guan, and T. T. Tjhung, "Full-rate full-diversity STBC with constellation rotation", *IEEE VTC-Spring 2003*, vol. 1, pp. 296–300.

[32] O. Tirkkonen, "Optimizing space-time block codes by constellation rotations", *Finnish Wireless Communications Workshop 2001,* pp. 59–60.

[33] N. Sharma and C. B. Papadias, "Improved quasi-orthogonal codes through constellation rotation", *IEEE Trans. on Communications*, vol. 51, pp. 332–335, Mar. 2003.

[34] D. N. Dao and C. Tellambura, "Optimal Rotations for Quasi-Orthogonal STBC with Two-Dimensional Constellations", *IEEE Globecom 2005*, pp. 2317–2321.

[35] A. Sezgin and E. A. Jorswieck, "On optimal constellations for quasi-orthogonal space-time codes", *IEEE ICASSP 2003*, pp. 345–348.

[36] A. Sezgin, E. A. Jorswieck and H. Boche, "Performance criteria analysis and further performance results for quasi-orthogonal space-time block codes", *IEEE ISSPIT 2003*, pp. 102–105.

[37] D. N. Dao and C. Tellambura, "A general method to decode ABBA quasi-orthogonal space-time block codes," *to appear in IEEE Communications Letters*, 2006.

[38] M. Y. Chen, C. Y. Chen, H. C. Li, S. C. Pei, and J. M. Cioffi, "Deriving new quasi-orthogonal space-time block codes and relaxed designing viewpoints with full transmit diversity", *IEEE ICC 2005*.

[39] M. Y. Chen, S. C. Pei, and H. J. Su, "Constellation expansion free quasi-orthogonal space-time block codes on square lattice constellations with full diversity and high coding gains", *IEEE PIMRC 2005*.

[40] C. Y. Chen, M. Y. Chen, J. M. Cioffi, "Full-diversity quasi-orthogonal space-time block codes for M-PSK modulations", *IEEE Globecom 2005*, pp. 3022–3026.

[41] M. Rupp, C. Mecklenbrauker, and G. Gritsch, "High diversity with simple space time block-codes and linear receivers", *IEEE Globecom 2003*, pp. 302–306.

[42] M. Rupp and C. F. Mecklenbrauker, "On extended Alamouti Schemes for space-time coding", *WPMC 2002*, pp. 115–119.

[43] L. He and H. Ge, "Fast maximum likelihood decoding of quasi-orthogonal codes", *Asilomar 2003*, pp. 1022–1026.

[44] A. Sezgin, E. A. Jorswieck, and E. Costa, "Lattice-reduction aided detection: spatial multiplexing versus quasi-orthogonal STBC", *IEEE VTC-Fall 2004*, pp. 2389–2393.

[45] C. Yuen, Y. L. Guan, and T. T. Tjhung, "Decoding of quasi-orthogonal space-time block code with noise whitening", *IEEE PIMRC 2003*, pp. 2166–2170.

[46] G. H. Golub and C. F. V. Loan, *Matrix Computations*, 3rd Edition, The Johns Hopkins University Press.

[47] C. Yuen, Y. L. Guan, and T. T. Tjhung, "Improved quasi-orthogonal STBC with group-constrained linear transformation", *IEEE Globecom 2004*, pp. 550–554.

[48] C. Yuen; Y. L. Guan; T. T. Tjhung, "Optimizing Quasi-Orthogonal STBC Through Group-Constrained Linear Transformation ", *accepted for publication in IEE Proc. Communications*.

[49] V. M. DaSilva and E. S. Sousa, "Fading-resistant modulation using several transmitter antennas", *IEEE Trans. on Communications*, vol. 45, pp. 1236–1244, Oct. 1997.

[50] N. Sharma and C. B. Papadias, "Full rate full diversity linear quasi-orthogonal space-time codes for any transmit antennas", *EURASIP Journal on Applied Signal Processing*, pp. 1246–1256, 2004.

[51] A. Boariu and D. M. Ionescu, "A class of non orthogonal rate-one space-time block codes with controlled interference", *IEEE Trans. on Wireless Communications*, vol. 2, pp. 270–276, Mar 2003.

[52] A. Yongacoglu and M. Siala, "Performance of diversity systems with 2 and 4 transmit antennas", *International Conference on Communication Technology 2000*, pp. 148–151.

[53] C. Yuen, Y. L. Guan, and T. T. Tjhung, "Algebraic relationship between quasi-orthogonal STBC with minimum decoding complexity and amicable orthogonal design", *IEEE ICC 2006*.

[54] Eric W. Weisstein., "Quaternion", From *Mathworld*, http://mathworld.wolfram.com/Quaternion.html.

[55] C. Yuen, Y. L. Guan, and T. T. Tjhung, "Construction of quasi-orthogonal STBC with minimum decoding complexity", *IEEE ISIT 2004*, pp. 309.

[56] C. Yuen, Y. L. Guan, and T. T. Tjhung, "Quasi-orthogonal STBC with minimum decoding complexity", *IEEE Trans. Wireless Comms.*, vol. 4, Sept. 2005, Pages: 2089–2094.

[57] C. Yuen, Y. L. Guan, and T. T. Tjhung, "Quasi-orthogonal STBC with minimum decoding complexity: further results", *IEEE WCNC 2005*, pp. 483–488.

[58] A. K. Zafar and B.S. Rajan, "Space-time block codes from co-ordinate interleaved orthogonal designs", *IEEE ISIT 2002*, pp. 275.

[59] R. A. Horn and C. R. Johnson, *Matrix Analysis*, Cambridge University Press, 1985.

[60] Z. A. Khan, B. S. Rajan, and M. H. Lee, "Rectangular co-ordinate interleaved orthogonal designs", *IEEE Globecom 2003*, pp. 2004–2009.

[61] Z. A. Khan and B. S. Rajan, "Single-Symbol Maximum Likelihood Decodable Linear STBCs", *IEEE Trans Information Theory*, vol: 52, pp. 2062–2091, May 2006.

[62] B. L. Hughes, "Differential space-time modulation", *IEEE Trans. on Information Theory*, vol. 46, pp. 2567–2578, Nov. 2000.

[63] G. Ganesan and P. Stoica, "Differential modulation using space-time block codes", *IEEE Signal Processing Letters*, vol. 9, pp. 57–60, Feb. 2002.

[64] Y. Jing and B. Hassibi, "Design of fully-diverse multi-antenna codes based on Sp(2)", *IEEE ICASSP 2003*, pp. 33–36.

[65] B. M. Hochwald and W. Sweldens, "Differential unitary space-time modulation", *IEEE Trans. on Communications*, vol. 48, pp. 2041–2052, Dec. 2000.

[66] B. M. Hochwald and T. L. Marzetta, "Unitary space-time modulation for multiple-antenna communications in Rayleigh flat fading", *IEEE Trans. on Information Theory*, vol. 46, pp. 543–564, Mar. 2000.

[67] B. M. Hochwald, T. L. Marzetta, T. J. Richardson, W. Sweldens, and R. Urbanke, "Systematic design of unitary space-time constellations", *IEEE Trans. on Information Theory*, vol. 46, pp. 1962–1973, Sept. 2000.

[68] B. L. Hughes, "Optimal space-time constellations from groups", *IEEE Trans. on Information Theory*, vol. 49, pp.401–410, Feb. 2003.

[69] V. Tarokh and H. Jafarkhani, "A differential detection scheme for transmit diversity", *IEEE Journal on Selected Areas in Communications*, vol. 18, pp. 1169–1174, July 2000.

[70] H. Jafarkhani and V. Tarokh, "Multiple transmit antenna differential detection from generalized orthogonal designs", *IEEE Transaction on Information Theory*, vol.47, pp. 2626–2631, Sept. 2001.

[71] M. Tao and R. S. Cheng, "Differential space-time block codes", *IEEE Globecom 2001*, pp. 1098–1102.

[72] Z. Chen, G. Zhu, J. Shen, and Y. Liu, "Differential space-time block codes from amicable orthogonal designs", *IEEE WCNC 2003*, pp. 768–772.

[73] C. Yuen, Y. L. Guan, and T. T. Tjhung, "Differential transmit diversity based on quasi-orthogonal space-time block code", *IEEE Globecom 2004*, pp. 545–549.

[74] C. Yuen, Y. L. Guan, and T. T. Tjhung, "Single-symbol decodable differential space-time modulation based on QO-STBC", *IEEE ICASSP 2005*, pp. 1069–1072.

[75] C. Yuen; Y. L. Guan; T. T. Tjhung, "Single–Symbol–Decodable Differential Space-Time Modulation Based on QO-STBC", *accepted for publication in IEEE Trans. Wireless Comms.*

[76] B. Hassibi and B. M. Hochwald, "Cayley differential unitary space-time codes", *IEEE Trans. on Information Theory*, vol. 48, pp. 1485–1503, June 2002.

[77] X. Xia, "Differentially en/decoded orthogonal space-time block codes with APSK signals", *IEEE Communications Letters*, vol. 6, pp. 150–152, April 2002.

[78] Y. Zhu and H. Jafarkhani, "Differential modulation based on quasi-orthogonal codes", *IEEE Trans. on Wireless Comms*, vol. 4, pp. 3005–3017, Nov. 2005.

[79] A. Hottinen, K. Kuchi, and O. Tirkkonen, "A Space-time Coding Concept For A Multi-Element Transmitter", *Canadian Workshop in IT 2001*.

[80] Z. A. Khan, B. S. Rajan, and M. H. Lee, "On Single-Symbol and Double-Symbol Decodable STBCs", *ISIT 2003*, pp. 127.

[81] B. S. Rajan, M. H. Lee, and Z. A. Khan, "A Rate-One Full-Diversity Quasi-Orthogonal Design for Eight Tx Antennas", *Internal Report TR-PME-2002-15*, available at: http://pal.ece.iisc.ernet.in/PAM/tech_rep02.html.

[82] C. Yuen, Y. L. Guan, and T. T. Tjhung, "A class of four-group quasi-orthogonal STBC achieving full rate and full divesity ", *IEEE PIMRC 2005*, pp. 92–96.

[83] C. Yuen, Y. L. Guan, and T. T. Tjhung, "On search of high-rate quasi-orthogonal space-time block code", *IEEE PIMRC 2004*, pp. 1647–1651.

[84] C. Yuen; Y. L. Guan; T. T. Tjhung, "On the Search for High-Rate Quasi-Orthogonal Space-Time Block Code", *accepted for publication in International Journal of Wireless Information Network (IJWIN)*, available at: http://dx.doi.org/10.1007/s10776-006-0033-2.

[85] U. Schendel, Sparse Matrices, numerical aspects with applications for scientists and engineers, John Wiley & Sons, 1989.

[86] S. Sahni, Data Structures, Algorithms, and Applications in C++, McGraw-Hill, 1998.

[87] C. Yuen, Y. L. Guan, and T. T. Tjhung, "New rate STBC with good dispersion property", *IEEE PIMRC 2005*, pp. 1683–1687.

[88] L .Liu and H. Jafarkhani, "Combining beamforming and quasi-orthogonal space-time block coding using channel mean feedback", *IEEE Globecom 2003*, pp. 1925–1930.

[89] B. Badic, P. Fuxjaeger, and H. Weinrichter, "Performance of quasi-orthogonal space-time code with antenna selection", *Electronic Letters*, vol. 40, pp. 1282–1283, Sept 2004.

[90] B. Badic, M. Herdin, H. Weinrichter, and M. Rupp, "Quasi-orthogonal space-time block codes on measured MIMO channels", *SympoTIC 2004*, pp. 17–20.

[91] A. Sezgin, E. A. Jorswieck, and E. Costa, "Optimal transmit strategies for QSTBC in MIMO Ricean channels with linear detection", *IEEE PIMRC 2005*.

[92] C. Toker, S. Lambotharan, J. A. Chambers, "Closed-loop quasi-orthogonal STBCs and their performance in multipath fading environments and when combined with turbo codes", *IEEE Transactions on Wireless Communications*, vol. 3, pp.1890–1896, Nov. 2004.

[93] J. K. Milleth, K. Giridhar, and D. Jalihal, "Closed-Loop Transmit Diversity Schemes for Five and Six Transmit Antennas", *IEEE Signal Processing Letters*, vol. 12, pp. 130–133, Feb 2005.

[94] Y. Yu, S. Keroueden, and J. Yuan, "Closed-Loop Extended Orthogonal Space–Time Block Codes for Three and Four Transmit Antennas", *IEEE Signal Processing Letters*, vol. 13, pp. 273–276, May 2006.

[95] N. Sharma, C. B. Papadias, "Reduced-complexity ML decoding of rate 6/8 and rate 1 linear complex space-time codes for up to eight transmit antennas with phase feedback", *IEEE Signal Processing Letters*, vol. 12, pp. 565–568, Aug 2005.

[96] A. Hottinen and O. Tirkkonen, "Matrix modulation and adaptive retransmission", *IEEE ISSSPA 2003*, pp. 221–224.

[97] C. K. Sung, J. Kim, and I. Lee, "Quasi-orthogonal STBC with iterative decoding in bit interleaved coded modulation", *IEEE VTC 2004*, pp. 1323–1327.

[98] J. Kim and I. Lee, "Space-time coded OFDM systems with four transmit antennas", *IEEE VTC 2004*, pp. 2434–2438.

[99] H. Jafarkhani and N. Seshadri, "Super-orthogonal space-time trellis codes", *IEEE Trans. on Information Theory*, vol: 49, pp. 937–950, April 2003.

[100]   H. Jafarkhani and N. Hassanpour, "Super-quasi-orthogonal space-time trellis codes", *IEEE ICC 2003*, pp. 2613–2617.

[101]   D. Wang, H. Wang, and X. Xia, "Space-time trellis code design based on super QOSTBC with minimum decoding complexity", *IEEE SPAWC 2005*, Page(s):161–165.

[102]   H. Mheidat, M. Uysal, and N. Al-Dhahir, "Time and frequency-domain equalization for quasi-orthogonal STBC over frequency-selective channels", *IEEE ICC 2004*, pp. 697–701.

[103]   L. He and H. Ge, "Quasi-orthogonal space-time block coded transceiver systems over frequency selective wireless fading channels", *IEEE ICASSP 2004*, pp. 321–324.

[104]   G. Yu and J. Kang, "Quasi-orthogonal space-frequency block code", *ISCIT 2004*, pp. 979–982.

[105]   X. Lu, G. Zhu, Y. Liu, X. Xiao, "A Rotation-Based Quasi-Orthogonal Space-Time Block Coded OFDM System over Frequency-Selective Channels", *IEEE CIT 2005*, pp. 485–489.

[106]   S. Gowrisankar and B. S. Rajan, "A rate-one full-diveristy low-complexity space-time-frequency block code (STFBC) for 4-tx MIMO-OFDM", *IEEE ISIT 2005*, pp. 2090–2094.

[107]   3Gpp – LTE, http://www.3gpp.org/Highlights/LTE/LTE.htm.

[108]   Motorola, "Cyclic shift diversity for E-UTRA DL control channels & TP", 3Gpp-LTE R1-060011, Jan 2006.

[109]   Samsung, "Performance comparison of EUTRA open loop transmit diversity techniques", 3Gpp-LTE R1-060814, March 2006.

[110]   Huawei, "Further results of transmit diversity schemes for common, distributed and broadcast channels", 3Gpp-LTE R1-060494, Feb 2006.

[111]   Nortel, "QO-STFBC / Double-STTD based H-ARQ for four transmit antennas", 3Gpp-LTE R1-060149, Jan 2006.

[112]   Nortel, "STBC based HARQ with simulation comparison", 3Gpp-LTE R1-060897, March 2006.

[113]   LG Electronics, "High-rate STC for open-loop MIMO with 2tx antennas", 3Gpp-LTE R1-060967, March 2006.

[114]   P. A. Regalia and S. K. Mitra, "Kronecker products, unitary matrices and signal processing applications", *SIAM Review*, vol. 31, pp. 586–613, Dec 1989.

INDEX